Travel Narratives, Travel Fictions

Travel Narratives, Travel Fictions

Daniel Cooper Alarcón

LEXINGTON BOOKS
Lanham • Boulder • New York • London

Published by Lexington Books
An imprint of The Rowman & Littlefield Publishing Group, Inc.
4501 Forbes Boulevard, Suite 200, Lanham, Maryland 20706
www.rowman.com

86-90 Paul Street, London EC2A 4NE

British Library Cataloguing in Publication Information Available

Library of Congress Cataloging-in-Publication Data

Names: Cooper Alarcón, Daniel, 1961- author.
Title: Travel narratives, travel fictions / Daniel Cooper Alarcón.
Description: Lanham, Maryland : Lexington Books, [2024] | Includes bibliographical references and index.
Identifiers: LCCN 2023051687 (print) | LCCN 2023051688 (ebook) | ISBN 9781666933772 (cloth : acid-free paper) | ISBN 9781666933789 (ebook)
Subjects: LCSH: Travel writing. | Travel in literature. | Tourism. | Population geography.
Classification: LCC G151 .C665 2024 (print) | LCC G151 (ebook) | DDC 910.4--dc23/eng/20240108
LC record available at https://lccn.loc.gov/2023051687
LC ebook record available at https://lccn.loc.gov/2023051688

™ The paper used in this publication meets the minimum requirements of American National Standard for Information Sciences—Permanence of Paper for Printed Library Materials, ANSI/NISO Z39.48-1992.

In loving memory of my beloved parents, Lynn and María. I miss you more than I can say.

Contents

Acknowledgments

This book has been both a labor of love and frustration that has taken me almost twenty-five years to bring to publication. It had its origins in the research for my previous book, *The Aztec Palimpsest*, in which I first explored travel narratives and travel fictions within the context of writing about Mexico and deepened my interest in tourism as a cultural practice. Around that same time (1995 or so), I began teaching a course at the University of Arizona I called "Theorizing Travel and Tourism," and since then I have taught various iterations of it regularly at both the undergraduate and graduate levels to the present day. This book would not have been possible without everything that I learned teaching that class over the intervening quarter century, and I owe a debt that I can never repay to all of the wonderful students who studied with me over the years and whose insights enliven this study in ways both large and small. To all my former students, many of whom I am still fortunate to be in touch with, thank you from the bottom of my heart.

In 1997, I had the opportunity to participate in a National Endowment for the Humanities (NEH) summer institute on the Maya world, where I made a life-long friend and became more interested in travel writing and the travel narratives of John Stephens, especially. I also had the privilege of studying with Mary Louise Pratt and Rolena Adorno, whose works have been enormously influential in shaping my own ideas about travel writing and transculturation. To Steve, Mary Louise, Rolena, and all the participants in that summer program, thank you—you made traveling around Central America for six weeks while I was deathly ill worth the effort.

I have many colleagues to thank as well. Susan Aiken and Larry Evers have been kind, generous, and wise mentors to me over the course of my career, and they remain cherished friends whom I admire more than I can say. Roger Bowen encouraged and supported the travel fictions project from its inception and was a gracious and generous reader of the first drafts of chapters, providing invaluable feedback on them. John Warnock, who shares an interest in travel literature, pointed me toward helpful and influential texts, as did

Suresh Raval. Roxanne Mountford and Bill Endres also supported my work and remain dear friends. Allison Dushane offered advice and encouragement over pints of beer (how I miss those get-togethers!), and Charlie Bertsch has been a kind and supportive friend and colleague ever since we met. And to my beloved friend Chris Carroll, I miss your larger-than-life presence and big heart every day. More recently, as I have struggled with personal losses and health issues, my colleagues have unfailingly come to my aid when I have needed help. In particular, I want to thank John Melillo, Lauren Mason, Paul Hurh, Ragini Srinivasan, Lee Medovoi, Lynda Zwinger, Tenney Nathanson, Adela Licona, Fred Kiefer, Jeff Banister, and Amy Fatzinger. And a very special thank you to my dear friend and colleague Aurelie Sheehan, whom I think of every day and who occupies a special place in my heart. There are no words to describe how much I miss you and how grateful I am for your friendship.

I also have many friends outside of the university to thank: Peter Lacy, Rebecca Rentería, Jim Gregory, Don Murray, Allie Baron, Elizabeth Menke, Liz Weibler, Jennifer Milbauer, Henrietta Barassi, and Dave Burkhalter have all supported and helped me time and time again over the years, as has my wise and dear friend Natasha. Doug Rosenberg, Steve Rice, John and Susan Neppl, John Riggins, Steve Brunner, Tom Gilgenbach, Bryan Peterson, Dean Quigley, Larry O'Brien, Bruce Cross, Chris Eide, Thomas Salmon, Alison Anderberg, and Wendy Walsh remain lifelong and cherished friends. There are doubtless many others whom I am overlooking, and to all of them, I beg their forgiveness and offer my undying thanks.

As time passed and life took its unexpected twists and turns the way it tends to do, work on the project stalled at various points. During this period, I was fortunate to have two mentorships that reinvigorated me and breathed new life into this book. The first was a three-week creative writing workshop with Carl Hiaasen, who encouraged me in my writing, both creative and academic. The second was an exciting and fulfilling NEH seminar on hemispheric American literature organized and led by Rachel Adams and Caroline Levander. I received invaluable feedback from them and from the other seminar participants on my John Stephens chapter, and the seminar is where I encountered *Volkswagen Blues* for the first time. My chapter on that novel owes an enormous debt to Rachel's analysis of it in her book *Continental Divides*, as well as to our seminar conversations. These opportunities to study with other writers and scholars for extended periods of time were consequential in allowing me to move forward in a productive manner on this project.

I also want to thank my patient and supportive editor, Holly Buchanan, for her guidance and help in bringing this book into print.

And, of course, I want to thank my beloved family members, to whom I owe everything. I love and miss my Mom and Dad more than I can say, and it is to them that this book is dedicated, in loving memory. I also love and miss my Tíos Pepe and Tavo and my Tía Ceci. And I am so grateful to still have my Tía Lety with us. To my *primos* and *primas* and *sobrinos* and *sobrinas*, thank you for everything. I thank my sister Susan and her family (a special shoutout to Stephen, Vincent, and Cecilia!) and my cousins Jim and Janice, all of whom have been very kind and supportive over the years. Thank you all from the bottom of my heart.

Last but certainly not least, I thank my best friend, Stephanie Athey. Thank you so much for all your wisdom, love, support, help, guidance, patience, and good humor all these years. A more enduring and fulfilling friendship I have never known.

Introduction

Part of the enduring appeal of travel narratives, a very old and durable genre, is the expectation of a window into exotic and unfamiliar worlds. Consider the following scene from Jack London's turn-of-the-century tramp memoir, *The Road* (1907):

> At other houses the doors were slammed in my face, cutting short my politely and humbly couched request for something to eat. At one house they did not open the door. I stood on the porch and knocked, and they looked out at me through the window. They even held one sturdy little boy aloft so that he could see over the shoulders of his elders the tramp who wasn't going to get anything to eat at their house. (16)

The family London describes metaphorically represents the contemporary readers of his memoir, who avidly read it for the view it offers into the mysterious world of the tramp and hobo. But, of course, windows work both ways and, as the passage clearly demonstrates, London uses his travel narrative to offer a subversive and unflattering critique of his curious, well-to-do readers. This is just one example of the many ways that accounts of travel perform important social and cultural criticism often overlooked in studies of the genre.

Keeping that in mind, in this scholarly investigation I have focused on a select and diverse array of texts we might broadly categorize as travel literature, to examine them not only in terms of the problems they create and perpetuate but also in terms of the critical insights they have to offer. I'm particularly interested in the relationship between nonfiction travel narratives and what I call travel fictions—novels and short stories that purposefully examine and critique different types of travel. While there has been, and continues to be, a substantial amount of scholarly work done regarding travel narratives, fictional accounts of travel have received far less critical attention, despite the fact that they have often anticipated and contributed ideas central to studies of travel, tourism, and migration.

I also use the term travel fictions to refer to the ways in which travel narratives (supposed factual accounts of travel) often misrepresent, distort, and fabricate notions about the people and places they purport to describe, creating useful mythologies, as well as to hint at the ways in which the acts of reading and writing have become an integral part of traveling—shaping not just itineraries but also perceptions and beliefs about the places travelers visit. Both types of travel writing are influential in this regard, and they often borrow from one another, mutually reinforcing ideas, tropes, and modes of representation, often in problematic ways. As David Johnson puts it, "Travel writing, then, is necessarily fiction, for the eye/I cannot see what the imagination, spurred by books, fails to conjure" (7). In other words, there is an interdependent relationship between travel narratives and travel fictions, one that produces durable and powerful mythologies of people, places, and cultures—with significant repercussions for the people being written about, as well as for the traveler and his or her culture. As Heidi Slettedahl Macpherson observes, "It is no surprise that travel texts perform the dual role of illumination and fictionalization, given that tourists and travelers themselves participate in a similar process of construction–of identity and nationality, self and other" (24). Thus, she continues, "To travel is to invent not only the destination, but also the traveler, as identity becomes a performance which tailors itself to whatever audience it encounters" (25). And, I would add, this invention takes place on multiple levels, both during the travels and again during the construction of the travel narrative.

Another focal point of this study is the different kinds of cultural work that travel literature performs at different historical moments. For example, travel narratives played a key role in promoting and sustaining imperial and colonial enterprises, and more recently, travel narratives and travel fictions have helped create an itinerary for travel to particular destinations and a set of criteria by which to evaluate a site's authenticity. A related phenomenon in the aftermath of World War II, and one reflected in the travel literature of the period, is the emergence of tourism as a type of travel deeply implicated in the formation of cultural identities and vital to the economies of many developing nations, as well as tourism's mirror image: the migration from developing parts of the world to more industrialized areas, driven usually by economic necessity or political conflict.

While it is important to address the ways in which travel narratives and travel fictions distort and misrepresent people, places, and cultures, it is my contention that both types of travel literature have also been used by many authors to theorize and think critically about travel and migration—in very sophisticated ways—an aspect of travel literature that has been neglected in much of the scholarship on the subject. Three studies which do pay close attention to the self-reflexive, critical, and theoretical sophistication of travel literature, and which have been particularly helpful as I have conducted my own investigation, are Marilyn Adler Papayanis's *Writing in the Margins: The*

Ethics of Expatriation from Lawrence to Ondaatje (2005), Macpherson's *Transatlantic Women's Writing* (2008), and Kai Mikkonen's *Narrative Paths: African Travel in Modern Fiction and Nonfiction* (2015).

In her study of expatriate literature, Papayanis argues that travel fictions do not inherently contribute to problematic mythologies of displacement, a position with which I wholeheartedly agree. Instead, she perceptively notes that, in many cases, they "actually call into question those mythologies and the modes of thought which inspire them" (4). She challenges, as I do throughout this study, "the notion that expatriate fiction is indifferent to the relations of inequality that inevitably attend—that enable—the practices it depicts by looking at the problem of self and Other," and she contends that "the degree to which certain expatriate narratives seriously and self-reflectively attend to issues of power and difference is startling" (7). Another aspect of Papayanis's examination of expatriate literature that has influenced my own is her recognition of the benefits of reading travel fictions alongside of and against travel narratives. As she puts it, when expatriate fictions are "read together with—or against—the individual writer's biography, correspondence, and self-presentation in travel writing and autobiography, a greater density of ethical thought emerges" (242).

Similarly, in her excellent study of late-twentieth-century transatlantic women's travel writing, Macpherson recognizes that both "factual" and fictional accounts of travel provide sophisticated explorations and critiques of gender, nation, and culture and their intersection. As she puts it, "Women writers have created a body of fictional and non-fictional narratives that appear to illuminate or mythologize aspects of national character, as well as to problematize the politics of tourism and travel in their negotiations of gendered identity" (24). Importantly, Macpherson insists upon treating travel fictions as texts with much to offer scholars investigating the complexities of transnational migration:

> Fictional travel narratives, like their 'real' counterparts, provide accounts of *difference* by exploring the politics of location while simultaneously examining localities. As cultural outsiders, these fictional travelers comment upon the society from which they spring and the societies to which they now belong. What all these fictional travel narratives have in common is a sense that gender is a condition which impinges upon—even determines—the travel experience. Each displays a postmodern sensibility, where identity is a performance which is fluid, not fixed. This is, of course, intimately bound up in travel itself. (26)

Mikkonen uses a similar approach in his study of travel narratives and travel fictions about sub-Saharan Africa, noting that cross-generic borrowing between the two forms was common, especially when the authors traveled with the hopes of gathering material for a book. Thus, these writers deploy novelistic devices in their travel narratives while drawing upon and re-imagining material from their travel notes and journals in creating their travel novels. He also discusses the reader's tendency to approach a nonfiction travel narrative with the expectation that it will provide an accurate representation of another place despite being organized by the traveler's mediating and subjective consciousness. Furthermore, although a nonfiction travel account may read as if it is being recorded from moment to moment, it is usually written after the journey is complete. And, like Macpherson before him, he notes that the travel writer writes in response to earlier accounts: "The close connection between travel writing and the reasons for travel, and the act of travelling [sic], is also observable in the writer's need to reconcile his or her experiences with the authority of earlier accounts of the same places and routes" (24). The cumulative effect of these earlier travel accounts is to create myths about places and cultures (often emphasizing the exotic or the primitive) so powerful that they exert tremendous influence over subsequent writers, resulting in a phenomenon in which "the fantasy of fiction . . . authenticates travel facts" (31).

While the comparative approach used by Papayanis, Macpherson, and Mikkonen has inspired and influenced my study, the work of numerous other scholars has also contributed to my critique of the representative travel narratives and fictions in the subsequent chapters. I discuss those scholarly contributions at length in the chapters that follow, but I would be remiss if I did not acknowledge how influential the work of Mary Louise Pratt and Rachel Adams has been in shaping my own approach to travel literature. I am very grateful to have had the opportunity to study with both while I was working on this project.

In closing this brief introduction, let me say a bit about the organization and approach used in this study. Each chapter focuses on a different type of travel, using representative texts to illustrate how travel narratives and travel fictions can both be used to theorize about and critique travel in sophisticated and innovative ways. In choosing those representative texts, I have elected to focus on the work of American writers in the broader hemispheric sense (the one outlier is Bharati Mukherjee, who was born in India but who emigrated to North America as a young woman and became an American citizen). I have also been mindful of both historical and cultural range: the works examined here run from the mid-nineteenth century to the late twentieth, and the authors come from diverse cultural backgrounds. Chapter 1 analyzes and discusses how the travel narratives and fictions written by Paul Bowles articulated and anticipated many of the ideas and questions that are now central to critical studies of travel, tourism, and migration. In this opening chapter, I also review the major scholarly works in that critical tradition. In Chapter 2, I focus on exploration, examining the famed travel narrative of John Stephens, describing his sojourn with Frederick Catherwood through Central America and southern Mexico in the mid-nineteenth century. In my analysis of his narrative, I show how Stephens's depiction of Latin America and the Maya sites he popularized shore up and promote an expansionist, nationalist US foreign policy while misrepresenting the people and places he describes in ways that would exert tremendous influence over later writers and the American reading public in shaping attitudes and beliefs about the region. Chapter 3 examines the emergence of the tramp and hobo in the post-Civil War United States through the exemplary tramp memoir of one of its most famous authors, Jack London. I illustrate how London's memoir, superficially a freewheeling adventure, provides a subversive critique of American capitalism directed toward London's most avid readers. Chapter 4 extends this focus on tramping by reading London's writings alongside a travel fiction that purposefully scrutinizes the tradition of travel writing that London and others belong to, even as it narrates its own road trip across the American continent. By juxtaposing the texts of London and Jacques Poulin, I show how travel narratives and travel fictions work synergistically, and I also demonstrate how travel fictions often self-reflexively critique the genre and tradition they belong to, even as they tell their own travel tale. In Chapter 5, I analyze and discuss Mukherjee's classic novel of immigration, *Jasmine*, which tracks a young Indian woman's journey and integration into US society during the 1980s—again, revealing the power of fiction to critically examine another type of travel in an insightful and provocative manner. The final two chapters reflect upon texts by Jamaica Kincaid, Clarice Lispector, and Lynne Tillman, with a focus on tourism and its impact upon less-developed nations, as well as the ways in which touristic discourse deploys popular tropes and myths taken from travel narratives and travel fictions, often reinforcing erroneous beliefs about people, places, and cultures. Ultimately, what I hope I have accomplished with this study is to illustrate the powerful and sophisticated critical work that both travel narratives and travel fictions are capable of and to persuade other scholars to begin to include travel fictions as part of their own investigations of travel and migration.

Chapter One

Travel Narratives, Travel Fictions

The Prescient Case of Paul Bowles

Paul Bowles was a prolific traveler, and his novels and short stories are set in those places he knew best, in particular North Africa and Latin America. But he also wrote memoirs—travel essays—about his journeys and stays in those regions, which he published in the collection *Their Heads Are Green and Their Hands Are Blue: Scenes from the Non-Christian World* (1963). These essays demonstrate two opposing tendencies that are also present in his earlier, famous first novel, *The Sheltering Sky* (1949): a deep affinity for a mythologized North Africa and, paradoxically, a deep interest in dismantling that mythology and in thinking critically about travel and tourism and their relationships to colonialism, nationalism, and cultural identity. It is this latter tendency that I think has been overlooked in much of the Bowles scholarship, and the oversight is not insignificant: In many ways, Bowles's critical thinking about travel and tourism anticipates by two decades the earliest theoretical work about various forms of travel.

Nevertheless, his travel essays about North Africa often romanticize the people, place, and cultures. For example, in "Africa Minor," Bowles writes:

> When I meet fellow Americans traveling about here in North Africa, I ask them, "What did you expect to find here?" Almost without exception, regardless of the way they express it, the answer, reduced to its simplest terms, is: a sense of mystery. They expect mystery, and they find it, since fortunately it is a quality difficult to extinguish all in a moment. They find it in the patterns of sunlight filtering through the latticework that covers the souks, in the unexpected turning and tunnels of the narrow streets, in the women whose features still go hidden beneath the *litham*, in the secretiveness of the architecture, which is such that even if the front door of a house is open it is impossible to see inside. If they listen as well as look, they find it too in the song the lone camel driver sings by his fire before dawn, in the calling of the muezzins at night, when their voices are like bright beams of sound piercing the silence, and, most often, in the dry

> beat of the *darbouka*, the hand drum played by the women everywhere, in the great city houses and in the humblest country hut. (23–24)

This is vintage Bowles, written in his clean, elegant style, and clearly reflects his affinity for a mythologized and romanticized North Africa. Throughout his travel essays about North Africa, he often dwells on the exotic—there are lengthy descriptions of religious cults engaged in rituals of bloodletting and hours of frenzied dancing resulting in trance-like states, as well as numerous tales of witchcraft and superstitions. The final essay in the collection describes a journey to a secret city hidden in a remote canyon where women are kept in harems as concubines. Perhaps not surprisingly, this tendency to represent North Africa with an emphasis on the exotic, sexual, and "primitive" is accompanied by language that sometimes slips into the colonial discourse of earlier eras. And, in one of the essays' most recurrent themes, Bowles speaks in wistful terms of the disappearance of traditional North African customs and lifestyles, even as he recognizes this change as an inevitable consequence of modernization. In all these regards, Bowles reproduces patterns—many of them problematic—familiar to readers of earlier travel narratives by Western authors about the non-Western and developing worlds.

Yet, paradoxically, Bowles carefully and deliberately sought to call the reader's attention to the ways in which stereotypes and myths have been constructed about North Africa, and he repeatedly shined a critical light on French colonialism, which he called "a reign of terror" (129). He voiced his support for the North African countries' self-governance, even though he feared the nationalist movements would destroy important indigenous traditions. And, perhaps, most interestingly, he very presciently predicted how tourism and nationalism were likely to impact one another in a post-colonial setting. In many ways, his trenchant observations anticipate those of Dean MacCannell in his book *The Tourist* (1976), often acknowledged as the first major scholarly work to focus upon tourism through a sophisticated theoretical and critical lens.

Like MacCannell, Bowles theorized that as the Western world became more modern and secularized, Westerners would increasingly travel to non-Western or developing nations in an attempt to reconnect with the sacred and the traditional: "In Tunisia, Algeria and Morocco there are still people whose lives proceed according to the ancient pattern of concord between God and man, agreement between theory and practice, identity of word and flesh (or however one prefers to conceive and define that pristine state of existence we intuitively feel we once enjoyed and now have lost)" (22).

Even if the Western tourist did not have such high-minded motivations, Bowles understood that those more traditional ways of living, precisely because they were becoming more scarce in the West, would exert a powerful attraction as they came to be seen as something quaint or picturesque: "The visitor to a place whose charm is a result of its backwardness is inclined to hope it will remain that way, regardless of how its inhabitants may feel. The seeker of the picturesque sees the spread of technology as an unalloyed abomination. Still, there are much worse things" (viii).

In addition, Bowles recognized that even as the Western traveler sought out this quaint, picturesque, traditional, and exotic foreign space—rooted in a North African mythology that was very much the product of varied discourses, including literature and his own literary contributions—such traditional customs and ways of living were rapidly giving way to more modern and Western ones. Bowles believed this was in part a product of years of European colonial domination but also a by-product of nationalism and a reaction against Western stereotypes regarding North Africa. As Bowles puts it:

> The attainment of political independence is only one facet of their problem. The North African knows that when it comes to appreciating his culture, the average tourist cannot go much closer toward understanding it than a certain condescending curiosity. He realizes that, at best, to the European he is merely picturesque. Therefore, he reasons, to be taken seriously he must cease being picturesque. Traditional customs, clothing and behavior must be replaced by something unequivocally European. In this he is fanatical. It does not occur to him that what he is rejecting is authentic and valid, and that what he is taking on is meaningless imitation. And if it did occur to him, it would not matter in the least. This total indifference to cultural heritage appears to be a necessary adjunct to the early stages of nationalism. (34)

The irony was not lost on Bowles. As he makes clear in the cited passage, he considered the turn away from tradition as, in part, a way for newly independent nations to reject the stereotypes that Westerners had made of them. But since these same newly independent nations, like Morocco, also recognized foreign tourists as an important source of external revenue, Bowles correctly predicted that the post-colonial nation would inevitably construct traditional and even stereotypical attractions, places, and spaces in order to fulfill the tourists' preconceived notions. In other words, exotic North Africa would be constructed and packaged for touristic consumption. In the following passages, Bowles articulates these trends in his usual perceptive and analytical style:

> On the whole, the new governments of Tunisia and Morocco wish to further tourism in their respective countries; they are learning that the average tourist is more interested in native dancing than in the new bus terminal, that he is more willing to spend money in the Casbah than to inspect new housing projects. For

> a while, after the demise of the violently unpopular Pasha of Marrakech, Thami el Glaoui, the great public square of Marrakech, the Djemaa el Fna, was used solely as a parking lot. Anyone will tell you that the biggest single attraction for tourists in all North Africa was the Djemaa el Fna in Marrakech. It was hard to find a moment of the day or night when tourists could not be found prowling around among its acrobats, singers, storytellers, snake charmers, dancers and medicine men. Without it Marrakech became just another Moroccan city. And so the Djemaa el Fna was reinstated, and now goes on more or less as before. (26)

A few pages later, he returns to this phenomenon:

> With Europeanization and nationalism have come a consciousness of identity and the awareness of that identity's commercial possibilities. From now on the North Africans, like the Mexicans, will control and exploit their own charms, rather than being placed on exhibit for us by their managers, and the result will be a very different thing from what it has been in the past. Tourist land it still is, and doubtless will continue to be for a while; and it is on that basis only that we as residents or intending visitors are now obliged to consider it. We now come here as paying guests of the inhabitants themselves rather than of their exploiters. Travel here is certain not to be so easy or so comfortable as before, and prices are many times higher than they were, but at least we meet the people on terms of equality, which is a healthier situation. (39)

Furthermore, like MacCannell after him, Bowles foresaw that as traditional activities were recast as performances for the benefit of tourists, they would become meaningless rituals, devoid of that deeper "sacred" quality which was at the core of what the tourist sought, whether he or she realized it or not. As Bowles puts it, "Even the works of man, unless they are being used in his daily living, have a way of losing their meaning, and take on the qualities of decoration" (vii).

Bowles's ambivalence toward the changes taking place in North Africa is perhaps best articulated in one of his later essays from the collection. In "The Route to Tassemsit," he describes his extensive travels across Morocco in search of traditional folk music, which he hoped to record and preserve for the Library of Congress. To assist with this project, he had been awarded a grant from the Rockefeller Foundation. The year was 1959. In one of the essay's most memorable episodes, Bowles recalls one of his contacts, a young Moroccan named Moulay Brahim, living in a sparse apartment in Marrakech. As Bowles describes him, Brahim is a young "deracinated" Moroccan, "militantly of his epoch," determined to break with traditional Moroccan values and customs. Brahim, Bowles tells us, sits in his spartan room for days at a time with his head bowed, not in prayer but in deep concentration as he listens to radio broadcasts from afar, sometimes with multiple transmissions overlapping. When Bowles suggests that he adjust the dial to focus on one frequency, Brahim explains that he likes the overlapping chatter because it reminds him of being in a café. Bowles concludes that for his friend, the radio broadcasts are neither a form of entertainment nor a medium of information, but rather a sort of "metaphysical umbilical cord . . . an essential adjunct to feeling that he is in contact with life" (164–65).

What a striking and revelatory contrast this episode provides: on the one hand, the American expatriate traveling about remote areas of Morocco, while attempting to record and preserve its folk music (his essays, of course, are another form of preservation) before it vanishes; and on the other, the young, modern Moroccan determined to leave those traditions behind, rarely leaving his small room as he listens continuously to radio broadcasts from the outside world, living in the presence of those foreign voices, a disembodied global community that provide him with more of a sense of belonging than the cafes of his own city.

Like his character Port in *The Sheltering Sky*, Bowles thought of himself as a traveler, not a tourist. In these essays, he distinguished between the former, those foreigners like himself who spend enough time in North Africa so that they can begin to learn its intricacies and perceive authentic culture, and the latter, those tourists who seek out a mythologized North African world—an exotic space comprised of stereotypes and distortions of elements of the traditional pre-colonial and colonial eras, a world that Bowles recognized would be increasingly manufactured for the tourist in order to benefit economically from those visitors. In the ironic epilogue to his essay "Baptism of Solitude," he alludes to the possibility of the Sahara itself becoming a profitable tourist attraction one day: "Since this piece was written, the Algerian war has changed the Saharan picture. Now the hypothetical voyager would probably not go back, because without special documents it is very unlikely that he would be allowed in. The Sahara is not on display at the present time" (144).

THEORIZING TRAVEL AND TOURISM IN *THE SHELTERING SKY*

Bowles's ideas about travel and tourism, discussed in the preceding section, represent an evolution of his thinking that began much earlier, with his first novel, *The Sheltering Sky* (1949). In what follows, I aim to lay out the prescient ways in which Bowles's mid-century novel anticipates much of the important, early theoretical work on travel and tourism by MacCannell (1976), Susan Sontag (1977), Paul Fussell (1980), Jonathan Culler (1981), and others. Bowles's novel is also a useful one with which to consider and "test" some of the ideas and theories of such major considerations of travel literature as Edward Said's *Orientalism* (1979), Mary Louise Pratt's *Imperial Eyes: Travel Writing and Transculturation* (1992), and Caren Kaplan's *Questions of Travel: Postmodern Discourses of Displacement* (1996). Before turning to Bowles, then, a few observations about these critical works, to set up some points of comparison.

In *Orientalism* and other works, Said documents and illustrates how a discursive network about the non-Western world is used by the West to represent, apprehend, manipulate, and "understand" its non-Western Others. Among these discursive strands are both travel narratives and the novel, two of the earliest and most durable vehicles for the West to represent non-Western cultures. And, in her influential study, *Imperial Eyes*, Pratt focuses on travel literature and traces its engagement with discourses of natural history, ethnography, empire, and capitalism—often in the service of those controlling enterprises. As a "travel fiction" by one of the most famous expatriate writers of the twentieth century, *The Sheltering Sky* needs to be considered as a literary work that is bound up with the literary traditions and conventions documented by Said, Pratt, and other scholars. Moreover, travel literature, and what I am calling travel fictions, have not only shaped ideas and perceptions about other places and cultures but also about the activity of travel itself, giving rise to what Kaplan has called "mythologized narrativizations [sic] of displacement" (2). Marilyn Adler Papayanis, in her study of expatriate writers, articulates some of the motifs common to such travel mythologies, particularly in the work of expatriate writers, including the desire to "escape from the world's increasing homogeneity, curiosity (sexual and otherwise) about 'Other' cultures, [and] the sufficiency of the expatriates's knowledge and technical proficiency to negotiate linguistic, cultural, and geographic difference" (156).

In *Questions of Travel*, Kaplan considers how contemporary literary and cultural criticism have contributed to such mythologies of displacement and expatriation. A good example of the phenomenon that Kaplan refers to is Paul Fussell's *Abroad* (1980), a discussion of English literary travelers during the 1920s and 1930s. In it, he writes that at the close of World War Two, "The British did awake, but to a different world, one in which the idea of literary traveling must seem quaint and a book about it a kind of elegy" (227). That different world—though he never says so explicitly—is a nascent post-colonial world. Fussell's lament intimates not only a mourning of something no longer possible but also a resentment toward those changes that make the kind of travel celebrated in his book not so much obsolete but democratized.[1]

It's not just that the formerly remote and exotic destinations of the privileged literary traveler are now the domain of the tourist, which bothers

Fussell, but also that there is another kind of travel, one "waiting in the wings" that threatens to replace the kind of romanticized travel that Fussell longs for. Tellingly, this new kind of travel surfaces only once in the book, obliquely, in a discussion of Evelyn Waugh, when Fussell suggests that "'in a world of displaced persons,' travel seems wholly anomalous" (215). Fussell's reference to displaced persons reminds us that since World War II, travel has not just radiated outward from metropolitan First World hubs, the former seats of European empire, but has increasingly flowed from the former colonies to the nations that once governed them. Also noteworthy are Fussell's attempts to distinguish between different kinds of travel. He asserts that, "Before tourism there was travel, and before travel there was exploration. Each is roughly assignable to its own age in modern history: exploration belongs to the Renaissance, travel to the bourgeois age, tourism to our proletarian moment. But there are obvious overlaps" (38). However, a few paragraphs later, his distinctions become more problematic:

> All three [types of travelers] make journeys, but the explorer seeks the undiscovered, the traveler that which has been discovered by the mind working in history, the tourist that which has been discovered by entrepreneurship and prepared for him by the arts of mass publicity. The genuine traveler is, or used to be, in the middle between the two extremes. If the explorer moves toward the risks of the formless and the unknown, the tourist moves toward the security of pure cliche. It is between these two poles that the traveler mediates, retaining all he can of the excitement of the unpredictable attaching to exploration, and fusing that with the pleasure of knowing where one is belonging to tourism. (39)

Reminding us that, etymologically, the word "travel" is derived from "travail," Fussell goes on to laud the traveler as one who combines intellectual inquiry with actual physical effort. But Fussell's claims that the traveler is motivated by a spirit of inquiry, while the tourist is merely purchasing the security of a cliché, are a bit misleading. The tourist is not without a sense of inquiry or curiosity—after all, don't both traveler and tourist seek out the extraordinary—or what Fussell calls the anomalous? Nor is the "literary traveler" any less predisposed to seek out or to project a predetermined quality onto the travel destination. What Fussell implies, but never quite says outright, is that a traveler's motives are different from and better than a tourist's. From Fussell's perspective, a traveler sacrifices or risks the body in the pursuit of the intellectual or literary, while a tourist cares only for physical pleasure and economic consumption. It, therefore, follows that a traveler must be more noble than a tourist and, since the traveler is driven by superior motives, his (and with Fussell, one must always use the male pronoun when speaking of these travelers) findings must be superior as well. Fussell's comment that a tourist purchases a cliché, an experience that, by definition, is trite, overused, and expected, implies that the traveler searches for something more significant, elusive, and unexpected. But what is that more significant something? Ironically, given Fussell's condemnation of tourism and tourists, a possible answer is provided by MacCannell in *The Tourist: A New Theory of the Leisure Class* (1976).

In this theoretical work, MacCannell speculates that tourism is a response to a pervasive fear that modern life is increasingly inauthentic or profane.[2] Thus, tourism is, above all, a quest for the authentic or sacred, driven by nostalgia, the increase in tourism paralleling the increasing anxiety of the modern individual that his or her reality is inauthentic. In this, MacCannell steers close to Baudrillard's concept of the "hyperreal" (there is no reality, just simulacra) and the effects of the anxiety produced by the hyperreal. Baudrillard tells us, "When the real is no longer what it used to be, nostalgia assumes its full meaning. There is a proliferation of myths of origin and signs of reality; of second-hand truth, objectivity, and authenticity" (171). MacCannell places tourist practices among this proliferation: "Sightseeing is a kind of collective striving for a transcendence of the modern totality, a way of attempting to overcome the discontinuity of modernity, of incorporating its fragments into a unified experience. Of course, it is doomed to eventual failure: even as it tries to construct totalities it celebrates differentiation" (13). Sontag advances a similar argument in her classic treatise *On Photography*: "People robbed of their past seem to make the most fervent picture takers, at home and abroad. Everyone who lives in an industrialized society is obliged gradually to give up the past, but in certain countries, such as the United States and Japan, the break with the past has been particularly traumatic" (10). This traumatic break with the past has been recognized by the tourist industry as an exploitable crisis and has resulted in a lucrative "nostalgia" or "heritage" industry that manufactures the past for the modern tourist, a past increasingly located in the developing world. The intense desire by the tourist to find something innocent, traditional, and, above all, authentic creates a situation in which those living in the developing world are coerced to live in the past or, as MacCannell suggests, are forced to enact a "backwardness" for tourists (178).

The term "backwardness" is loaded with suggestive implications that throw a spotlight onto what the modern tourist hopes to gain from such (post-)colonial encounters. From the tourist's point of view, the developing world is backward in ways that buttress the tourist's sense of superiority: technologically, educationally, and economically. But the natives of these countries are backward in the temporal sense as well: they are staged back in time, a removal which grants them an aura of innocence and authenticity, and it is this removal which enables the only superior position the native is permitted: a prelapsarian state of grace. But, following MacCannell's theory, it is all an illusion, an act, an economic "survival strategy" that provides a "cultural service stop" in exchange for tourist dollars. This "forced traditionalism" or enactment of "backwardness" in all of its many connotations is one form of what MacCannell calls "staged authenticity" (178).

At first glance, "staged authenticity" appears to be an oxymoron, but the term is more accurately a useful redundancy: authenticity is always bestowed, invested, marked, adjudicated, certified, framed, and, therefore, always staged in some fashion. Moreover, authenticity is extremely fragile, primarily because the process of authentication depends upon that of replication. Drawing on the ideas of Walter Benjamin in his famous essay, "The Work of Art in the Age of Mechanical Reproduction," MacCannell goes so far as to argue that a tourist site/sight becomes authentic only after the first copy of it is produced (MacCannell 47–48). This creates a dilemma neatly articulated by Culler in an oft-quoted passage: "The paradox, the dilemma of authenticity, is that to be experienced as authentic it must be marked as authentic, but when it is marked as authentic it is mediated, a sign of itself and hence not authentic in the sense of unspoiled" (Culler 137). If MacCannell and Culler are right, there is more than a little irony in the suggestion that the tourist seeks the authentic out of a fear that his/her world is inauthentic, while tourist sites can be invested with authenticity only by replicating and marking them, replicas and signs which increase the sense that the world is inauthentic. Or, as Sontag puts it, "It is reality which is scrutinized, and evaluated, for its fidelity to photographs" (87), a fidelity that is increasingly difficult to achieve. Furthermore, the mere presence of a tourist has the potential to undermine a site's authenticity, an idea that can perhaps best be suggested by reworking a familiar conundrum: *If a tree falls in a forest and a tourist is present, does it make an authentic sound?* Fussell, I suppose, might respond by stating only if a traveler is present because it is the traveler, not the tourist, who is intellectually equipped to discern the authentic.

Of course, one obvious objection to these theories is that their near-exclusive focus on tourism as a First World phenomenon overlooks other types of travel originating within the developing world, including migration and mass dislocations caused by war, famine, drought, and poverty. In *Questions of Travel*, Kaplan addresses this oversight in much of the early critical work done on travel and tourism, arguing that as scholarship began to account for it, new terminology was deployed by critics, and certain kinds of movement from place to place—for example, exile, diaspora, and deterritorialization—acquired a privileged status that went largely unchallenged. As Kaplan astutely points out, attempts to define and distinguish these different types of movement from one another are as tricky and problematic as Fussell's attempts to distinguish between exploration, travel, and tourism.

Making a similar argument, Papayanis has suggested that as scholars became more critical of certain types of expatriate literature as complicit in furthering and supporting a colonial discourse which disenfranchised, primitivized, and subordinated non-Western subjects, they began to stop asking whether those travel fictions could also effectively critique colonial attitudes and ideas. She argues, as I do here, that many expatriate fictions, including *The Sheltering Sky*, "actually call into question those mythologies [of displacement] and the modes of thought which inspire them," and she argues for a more nuanced reading of expatriate texts (4). As she puts it, "Indeed, the degree to which certain expatriate narratives seriously and self-reflexively attend to issues of power and difference is startling" (7). But rather than focus on the undoing of the Western self in its encounter with alterity, as Papayanis and other critics frequently do when discussing *The Sheltering Sky*, I wish to consider the ways in which Bowles uses fiction to formulate sophisticated and critical ideas about different types of travel and their relationship to colonial power and privilege in the immediate aftermath of World War II—ideas that anticipate with almost uncanny presentiment some of the major theories just discussed.[3] In what follows, I argue that *The Sheltering Sky* is a travel fiction that energetically and self-consciously engages many of the ideas central to later studies of colonial discourse, travel literature, tourism, and other types of travel.

THE SHELTERING SKY

> He did not think of himself as a tourist; he was a traveler. The difference is partly one of time, he would explain. Whereas the tourist generally hurries back home at the end of a few weeks or months, the traveler, belonging no more to one place than to the next, moves slowly over periods of years, from one part of the earth to another. (Bowles 13–14)

This distinction between traveler and tourist—made early in *The Sheltering Sky*—is the first of several moments in which Bowles anticipates many of the ideas which have become central to scholarly studies of travel and tourism, long after the novel's 1949 publication. Like Fussell, Bowles's protagonists, Port and Kit Moresby, value travel as a form of intellectual inquiry far superior to tourism and view the effects produced by World War II as a setback to such travel.[4] As the novel opens, Bowles's narrator informs the reader that Port and Kit have arrived in North Africa "with the intention of keeping as far as possible from the places which had been touched by the war. For, as [Port] claimed, another important difference between tourist and traveler is that the former accepts his own civilization without question; not so the traveler, who compares it with the others and rejects those elements he finds not to his liking. And the war was one facet of the mechanized age he wanted to forget" (14). Like Port, many of the British writers discussed by Fussell fled their homeland to escape the horrors of the "mechanized age" and, in particular, mechanized warfare and what it had wrought. And like Port, these writer-travelers rejected much of their own culture as inferior and limiting and sought renewal in other cultural forms and ideas.[5]

Port's fascination with so-called exotic places is established early in the novel. His name, Port Moresby, is also the name of a harbor in New Guinea; he is fascinated with maps, and, perhaps most importantly, he has come to North Africa in search of someplace untouched by the Second World War. Again anticipating another of Fussell's main contentions, Bowles suggests that World War II has ruined travel by disrupting distinct cultural traditions and triggering a kind of massive, global, cultural homogenization.[6] Kit comments, "The people of each country get more like the people of every other country. They have no character, no beauty, no ideals, no culture. Nothing, nothing" (16). Port concurs: "You're right. . . . Everything's gray and it'll be grayer. But some places will withstand the malady longer than you think. You'll see, in the Sahara here" (16). Greg Mullins perceptively observes that Bowles uses this exchange between his two principal characters to remind the reader that "Port's romance with premodernity is a fantasy of his own (or of his culture's) construction" (24).

Port's suggestion that something culturally authentic still exists in remote parts of the world dovetails neatly with MacCannell's theory that what the First World tourist or traveler seeks is an idealized, pre-industrial world of tradition and stability, increasingly thought to be located in developing countries. As noted earlier, MacCannell theorized that tourism is a response to a pervasive fear that modern life is increasingly inauthentic or profane. In his study of Bowles's fiction, Richard Patteson makes a similar claim about modernity. He observes that the twentieth century has been marked by wars, the realignment of national boundaries, mass migrations across the globe, and the

> collapse of traditional customs as newly independent governments rush to modernize. But there is a larger dimension, too. The long decline of faith throughout the Christian west and the increasing secularization of life nearly everywhere have contributed to an intuition of exposure and nakedness that transcends politics. And it is this perception of ultimate homelessness that Paul Bowles, expatriate writer, most powerfully images in his fiction. (x–xi)[7]

In his analysis of the contemporary Anglophone travel novel, including *The Sheltering Sky*, Stephen Levin advances a similar view, noting that these travel fictions frequently "portray a subject striving to come into being in the face of radical alienation" and searching for a "utopian alternative to the myriad, and always tainted, subject-positions of modernity" (142–43).

MacCannell, Patteson, and Levin contend that these global events and changes have created a tremendous sense of loss in the modern individual, which, in turn, generates a quest for something meaningful to fill the void. Such anxiety is ever present in *The Sheltering Sky*. As Port says to Kit during a moment of reconciliation between the estranged couple, "I think we're both afraid of the same thing. And for the same reason. We've never managed, either one of us, to get all the way into life. We're hanging on to the outside for all we're worth, convinced we're going to fall off at the next bump. Isn't that true?" Frightened, Kit can only respond, "But if we're not in, then we are more likely to fall off" (101–2).

Port's uncertainty about his existence and grounding in some authentic reality manifests itself in other ways. In an exchange with his American friend, Tunner, Port fiercely rejects the notion that a person's existence is worthwhile only if he or she demonstrates a concern for others:

> I don't have to justify my existence by any such primitive means. The fact that I breathe is my justification. If humanity doesn't consider that a justification, it can do what it likes to me. I'm not going to carry a passport to existence around with me, to prove I have the right to be here! I'm here! I'm in the world! But my world's not humanity's world. It's the world as I see it. (95)

Such a defensive response suggests that this self-professed loner is not as self-sufficient as he claims, and it soon becomes clear that Port often requires Kit's presence to authenticate his experiences or to heighten his sense of reality. Even when she is not present, he summons her with his imagination, as in the following passage, describing a solitary walk he takes early in the novel:

> A faint vision began to haunt his mind. It was Kit, seated by the open window, filing her nails and looking out over the town. And as he found his fancy returning more often as the minutes went by, to that scene, unconsciously he felt himself the protagonist, Kit the spectator. The validity of his existence at that moment was predicated on the assumption that she had not moved, but was still sitting there. It was as if she could still see him from the window, tiny and far away as he was, walking rhythmically uphill and down, through light and shadow; it was as if only she knew when he would turn around and walk the other way. (24)

Like Sontag's tourists who measure reality against photographs and replicas and often find it lacking, Port can only enjoy his walk through the Algerian city by conceiving of it as some sort of spectacle in which he is an actor.

Moreover, like Culler's tourist who becomes suspicious about the authenticity of a sight when no visible markers are present, Port requires Kit's presence to authenticate his experience (and even his very existence). We can note a similar anxiety in a later episode, in which Port takes Kit with him to gaze upon a vast desert landscape, hoping that her presence will authenticate the sight for him, only to return to the same spot later that evening, without her, as though her presence earlier had deprived him of something (98–103). This paradoxical behavior on Port's part is completely in line with Culler's dilemma of authenticity: Port requires Kit's presence to validate the experience, and yet her very presence prevents him from having an unmediated encounter with the site, prompting him to return without her. The dilemma becomes even more acute later in the novel, as Port reflects that the very act of perception appears to taint or spoil his ability to apprehend "pure" experience: "The landscape was there, and more than ever he felt he could not reach it. The rocks and the sky were everywhere, ready to absolve him, but as always he carried the obstacle within him. He would have said that as he looked at them, the rocks and the sky ceased being themselves, that in the act of passing into his consciousness, they became impure" (168).

TEA IN THE SAHARA

Bowles designed his novel in three parts, or books, as he called them. "Tea in the Sahara," the title of Book One, resonates throughout the entire novel. It refers to, among other things, a story told to Port by Smail, an Arab man whom he meets during his aforementioned walk. Smail arranges a meeting for Port with a young woman named Marhnia. When Port inquires whether she is a prostitute, Smail angrily objects, insisting that she is a dancer; this is much in the same way that Port insists that he is a traveler, not a tourist. The encounter with Marhnia is notable for the way in which both Port and Marhnia try to manipulate one another in a manner that clearly reveals the antagonistic nature of the tourist/host relationship. It also, of course, engages Orientalist discourse and takes up that tradition's tendency to represent the Orient as a sexual playground for Western men. Throughout the novel, Port repeatedly counts on his economic power to coerce the natives into giving him whatever he wants, including sex. But before she allows the sexual encounter, Marhnia insists Port listen to a story, a parable about three Arab dancing girls whose dream is to have tea in the Sahara. Hopelessly in love with a handsome stranger and feeling that life is passing them by, the three women spend their life savings traveling with a caravan into the desert, where they hope to have tea. Leaving the caravan, they climb dune after dune, in search of the highest one. When they finally ascend the highest dune and fall asleep, exhausted, they die in the midday heat, their teacups full of sand.

Although the little parable foreshadows Kit's experience later in the book, it also functions as an allegory for the entire novel. Like the women in Marhnia's story, Port, feeling that life is eluding him, plunges recklessly farther and farther into the Sahara, never satisfied with any place he comes to, only to die deep in the interior of the desert.[8] The teacups full of sand suggest the fragility and incongruity of civilized ritual in the wilderness and how ill-equipped the travelers in each narrative are for the wilderness into which they journey (Port and Kit's traveling companion, Tunner, for example, recalls that Port, who dies of typhoid fever, never bothered to get vaccinated prior to the trip). In addition, tea is an ancient drink that originated in the Orient and then, later, as European nations colonized the East, became a popular imported beverage, especially for the British. Therefore, tea also symbolizes the long historical conflict between East and West. We might also think about the setting of both the parable and Bowles's novel in the Sahara: a seemingly desolate place filled with oases, a characteristic which serves nicely as a metaphor for the emotional desolation of Port and Kit and their hopes for a reconciliation in the desert. The Sahara is also highly unstable, its massive sand dunes shifting continually with the powerful winds that sweep across it, and this characteristic instability also works well to mirror the thematic concerns of the novel, reflecting both Port and Kit's anxieties regarding a perceived lack of stability in the world around them and in their own respective identities.

Impatient with the story and the storytelling ritual, Port tries to hasten along the proceedings and, after the quick sexual encounter is completed, catches Marhnia trying to steal his wallet, once again underscoring the economic and antagonistic basis of tourist/host relations. The episode ends with Port fleeing the encampment, barely escaping its angry inhabitants, and is the first of many that highlight Port's ambivalence toward the pre-industrial, non-Western world he claims to want to know so badly, and that he journeys into during the course of the novel.[9] Ironically, the war, which early in the narrative had been blamed for disrupting distinct cultural traditions, is also credited with utterly destroying the tourist infrastructure in the interior of French North Africa, much to the dismay of the American travelers. The ambivalence also attaches itself to the very act of travel. On the one hand, Port finds that he is able to think more clearly and objectively about problems while traveling—the act of travel creates a space for introspection and reflection (105), and he is excited by the thought that he is venturing into the unknown, at one point indulging in the fantasy that he is a pioneer, exploring the frontier as he imagines his great-grandparents did (108). On the other hand, he also worries about personal safety as official government bulletins warn against traveling into the interior of North Africa (108). As the Americans move farther south into ever more remote areas of the Sahara, the complaints about cultural homogenization produced by globalization and mechanization are replaced by complaints about the lack of tourist infrastructure, revealing the American tourists' inability to adapt to the cultural differences they so longed to experience. As the journey into the interior continues, the Western travelers, including Port, become increasingly obsessed with the quality of their food and lodging. As the hotels, food, and transportation become less Western, less opulent, less comfortable, the dissatisfaction of the travelers grows. In a memorable scene, after having traveled deep into Algeria, Port enters his hotel room to find that Kit has emptied out the contents of her suitcases, a "pathetic little fortress of Western culture," and is looking at her belongings in an effort to comfort herself (162). As she puts it, "When I looked out that window after lunch, I felt I'd simply die if I didn't see something civilized soon" (161). Despite the desire for something real and meaningful that they hope they can find in the desert, the two principal characters at this point are still operating under the civilized/primitive opposition so central to Western discourse about the Orient and Western tourism in the developing world. If whatever elusive truth they seek is *out there*, in the untamed desert wilderness, it is also clear that they are fearful of its destabilizing effects. That their fears are well-founded becomes clear in the disturbing final third of the novel, which deals with Kit's decision, following Port's death, to venture even farther into the Sahara.

RELATIONS WITH NORTH AFRICANS

Both Port and Kit regard the indigenous North Africans in the most simplistic of terms, but Bowles makes the reader keenly aware that the Americans are as much objects of curiosity for the natives as they are for the Americans, and in several scenes, he reverses the colonial gaze. For example, upon arrival in Algeria, Port is unnerved by the stares and murmurs that his presence generates from the local inhabitants, which cause him to wonder, "How friendly are they? Their faces are masks. They all look a thousand years old. What little energy they have is only the blind mass desire to live, since no one of them eats enough to give him his own personal force. But what do they think of me? Probably nothing" (22). (Kit also feels uncomfortable as a foreign presence and wishes they had gone to Europe, "where one was not inevitably on display every time one stirred out of the hotel" [176–77].) Shortly after this passage, Port meets Marhnia. The sexual encounter with her will be the first instance of Port's desire to use Arab women for sex and is echoed by a subsequent episode in which he searches in vain for a blind girl he wants to have sex with at a brothel. In the first instance, Marhnia requires Port to listen to her story of "Tea in the Sahara" as a condition for having sex and, in the latter instance, Port deals with his sexual frustration at not being able to locate the blind girl by bribing an old woman to make him a pot of tea. Port's Arab companion Mohammed assumes that by "having tea," Port means he had sex with the old woman to compensate for not finding the blind girl: "Two hundred francs for tea! Too much for such old tea! I hope you had two glasses my friend" (143–44). As these episodes illustrate, Port counts on his economic power to get what he wants from the natives, and as is the case for most tourists, it is primarily at the economic level that Port interacts with the local people. But, in Marhnia's and the old woman's negotiations and stipulations, Bowles provides good examples of Pratt's contention that a colonized people are not without a certain degree of control over the rules imposed upon them.[10]

While Port is guided by his selfish desires, his perception of the natives is that they are untrustworthy, simple-minded, and greedy. Similarly, Tunner, despite developing a friendship with the innkeeper Adelkader, reacts with disgust to the natives he encounters in the streets: "There were days when he felt contempt for these absurd people; they were unreal, not to be counted seriously among the earth's inhabitants" (257). And all three Americans display the common touristic tendency to view the native inhabitants as existing in the past. At one point, Kit gazes upon a Saharan village and wonders if anything like it could exist in the "Atomic Age" (185). When she objects to Port's plan to usurp the place of two Algerians on a sold-out bus, Port replies, "Time doesn't exist for them" (177).

But of the three Americans, it is Kit who has the most intimate and involved dealings with the natives, and it is here that Bowles draws on the rich depository of Orientalist imagery, creating lurid and horrific portraits.[11] In an early chapter, Kit flees Tunner's sexual advances on a train by slipping away and entering the fourth-class cabin. What she encounters there is presented by Bowles as a chamber of horrors: people are sandwiched into the train car like sardines in a tin, a man calmly eats a snack of red locusts, another holds a severed sheep's head in his hands, and the scene ends with Kit gazing into "the most hideous human face she had ever seen"—a tall, hooded man missing his nose (84–85).

The depiction of the natives presented in these early scenes in the novel establishes them as primitive beings, a pattern reinforced in the disturbing, final section of the book, as Kit literally enacts Marhnia's parable, joining a desert caravan where she becomes the sexual captive of a traveling merchant. It is in this third, final section of the novel that Bowles gives full power to the Orientalism that had been operating as an undercurrent throughout the first two-thirds. Kit's captivity at the hands of Belquassim taps into the long history of European literature about the Arab world so carefully traced by Said in *Orientalism*. For example, one of Kit's guards is described as a large African slave with "elephantine skin" and another is a diminutive boy named Othman, described as "frog-faced" and "reptilian." As Said points out, the dehumanization in these descriptions is a common feature of Orientalist texts. Raped repeatedly, drugged, and imprisoned for days in a small cell, Kit herself is dehumanized and reduced to a primitive state. Although she does not die a physical death, her captivity deprives her of memory, language, and, ultimately, her identity. As her name suggests, Kit is closely associated with her ever-diminishing luggage, which carefully registers the gradual dissolution of her identity as she plunges deeper into the interior after Port's death until, by novel's end, she is no longer Katherine Moresby but instead an object that is carefully guarded and returned to the American embassy by a series of mechanics and pilots who are careful never to let go of her. Ironically, by the end of the novel, the only belonging that Kit has been able to retain is her passport, which, at this point, is wholly meaningless to her. While it may be true, as Patteson suggests, that a recurring theme in Bowles's work is the refusal of "alien cultures" (non-Western) to be domesticated by outsiders (Westerners), one must also note, as Allen Hibbard does, that such resistance is often presented by Bowles in the violent, primitive terms that form one strand of Orientalist discourse.[12] Or, as Papayanis puts it, Kit's horrifying destruction of self should not blind us to the degree to which "the Other is there to facilitate various forms of self-dismantling rather than to become a partner in dialogue" (xii).

It is also important to recognize that Bowles takes great pains to suggest that both Port and Kit carry the primitive within them, that it is a basic part of human nature, and that what we might call civilization is merely a façade. In a 1981 interview in the *Paris Review*, Bowles remarked, "If I'm persuaded that our life is predicated upon violence, that the entire structure of what we call civilization, the scaffolding that we've built up over the millennia, can collapse at any moment, then whatever I write is going to be affected by that assumption" (as quoted in Patteson xi).[13] If Port's primitive nature is most often revealed in his tendency to sexualize the North African women, we are told early in the novel that Kit is continually torn between intellect and instinct: "One reason [Kit] had such a strong dislike of hearing dreams recounted was that the telling of them brought straightway to her attention the struggle that raged in her—the war between reason and atavism" (44). This internal conflict causes Kit to rely on Port to make decisions for her, and she is often depicted as sitting passively, looking out of windows. But with Port's death, a sense of empowerment overtakes Kit, and she determines to hold onto that freedom:

> Life was suddenly there, she was in it, not looking through the window at it. The dignity that came from feeling a part of its power and grandeur, that was a familiar sensation, but it was years ago that she had last known it. . . . Swiftly she walked along, focusing her mind on that feeling of solid delight she had recaptured. She had always known it was there, just behind things, but long ago she had accepted not having it as a natural condition of life. Because she had found it again, the joy of being, she said to herself that she would hang on to it no matter what the effort entailed. (247–48)

It is at this point that Kit joins Belquassim's caravan, and her decision to do so heightens her sense of empowerment: "Even as she saw these two men she knew that she would accompany them, and the certainty gave her an unexpected sense of power: instead of feeling the omens, she now would make them, *be* them herself" (262–63).

Bowles frames this final third of the novel with the following epigraph from Franz Kafka:

"From a certain point onward there is no longer any turning back. That is the point that must be reached." Kit clearly ventures beyond this point in her effort to empower herself and her experiment in freedom, sadly, is doomed by it. Kit's loss of identity begins with her making a concerted effort to repress memories of Port's death and of her former life: "Resolutely she turned her mind away, refusing to examine it, bending all her efforts to putting a sure barrier between herself and it." (267)

Following the trauma of repeated rape at the hands of Belquassim, Kit begins to lose her mind and becomes an instinctual being: "It was so long since she had canalized her thoughts by speaking aloud, and she had grown accustomed to acting without the consciousness of being in the act. She did only the things she found herself already doing" (276). After her long captivity, Kit escapes and is returned to the American consulate, only to flee once again rather than be forced to remember and return to her former identity. Knowing that Bowles was an admirer of Edgar Allan Poe, I have often thought that what makes *The Sheltering Sky* so terrifying, in a manner reminiscent of Poe, is how comfortable Kit becomes in her atavistic identity. Like the Professor in "A Distant Episode," the Bowles short story that was a precursor to *The Sheltering Sky*, Kit is frightened by the thought of having to confront her previous self.[14] Timothy Weiss astutely observes:

> Bowles's narratives of voyage de-romanticize the traveler and dismantle the modern belief in the individual as an autonomous self; in these stories, published during a period of the dissolution of European empires in the late 1940s and early 1950s, the imperial-colonial equation, which often worked to privilege and protect nineteenth-century European voyagers, breaks down, revealing the Western traveler's vulnerability in alien contexts. The barrier separating

> European and non-European, colonizer and colonized, weakens, no longer demarcating clear, unquestioned, unambiguous identities. Voyages split the Western self from its cultural, social base and set it adrift in another space where anything can occur. . . . When the individual—in Bowles's stories, the Westerner especially—becomes cut off from his cultural, social context, a possibility that travel in a foreign land always presents, he becomes disoriented and capable of a range of mimetic transformations. The ego-centered point of view of the Western 'I' is dismantled and shown in all its vulnerabilities. (55)

If, as the Kafka epigraph suggests, to strive for some metaphysical point of no return is a worthwhile goal, Bowles seems to warn, through the respective fates of both Port and Kit, that reaching and venturing beyond that point carries an extremely high price and one that is nonrefundable. In Kit's case, I cannot help but note that her movement past the point of no return is a movement backward, a retreat into that primitive human nature we all carry within us.

PASSPORT TO EXISTENCE

Readers of the novel will likely recall the anxiety caused to Port by the theft of his passport, contrasted here by Kit's indifference to her own. On the one hand, Port feels exhilarated by the prospect of traveling into the Sahara with no proof of his identity. On the other, the loss of his passport makes him feel only half-alive. Port's mixed feelings about his lost passport place him alongside of those British writers studied by Fussell, who found passports a nuisance. Fussell points to the institutionalization of the passport, and reactions to it, as one of the most visible markers of the insecurity and anxiety characteristic of the modernist period, a characteristic that is also a distinguishing feature of Bowles's work. As Fussell puts it:

> So small a phenomenon as the passport picture is an example of something tiny which has powerfully affected the modern sensibility, assisting that anxious self awareness, that secret but overriding self contempt, which we recognize as attaching uniquely to the world of Prufrock and Joseph K. and Malone. There are other unprecedented contributors [sic] to the modern neurosis: things like the numbers used in the modern world for personnel [sic] identification and coercion: social security numbers, taxpayer numbers, driver license numbers, license plate numbers and, on the Continent, identity card numbers, But the passport picture is perhaps the most egregious little modernism. (26)

Passports are clearly important symbols in this literary work. Port's name, while more obviously a reference to Port Moresby, New Guinea, also suggests a link to the government document authenticating one's identity. As noted earlier, Port struggles mightily with feelings of insecurity and, in the argument with Tunner, angrily rejects the notion that the individual requires some sort of recognition from others—what he calls a passport to existence—to legitimize one's identity. Despite his proclamations to Tunner, however, Port's deep-seated insecurities over his identity, his place in the world, and even his own existence intensify as he travels deeper into North Africa. Significantly, it is the theft of his passport that exacerbates many of these insecurities. Reporting the theft of his passport to the French authorities, Port tells Lieutenant d'Armagnac, "It's strange how, ever since I discovered that my passport was gone, I've felt only half-alive. But it's a very depressing thing in a place like this to have no proof of who you are, you know" (160). It's the phrase "in a place like this" that I find so tantalizingly suggestive. Does Port mean in a foreign country where one requires a passport to enter or exit? Or does he mean in a European colony, where American citizenship would presumably ensure access to privileged areas? Or does he mean in a remote area, where, without an official document, one's identity might be questioned? Through Port's comments, we begin to get a sense of how the passport functions not only as a marker of nationality, to grant certain privileges, but also to authenticate one's very identity and existence—to take something inherently fluid and fix it in time and space.

Whatever Port means, it seems clear that, on some level, he realizes the risks of continuing to travel without the passport. And yet, this is precisely what he chooses to do, motivated primarily by his desire to keep Kit away from Tunner, who is returning his passport to him. Paradoxically, although he tells the Lieutenant that he feels only half alive without it, he also feels exhilarated by the thought of traveling without it: "When he considered it, he realized now that it rather suited his fancy to be going off with no proof of his identity to a hidden desert town about which no one could tell him anything" (169). As Rachel Egboro has remarked, for Port, the passport is a type of shelter; it secures his place even when he is out of place, and he is exhilarated to abandon one form of shelter to try his luck and his theory of no passport to existence, relying instead upon the unpredictable shelter of the sky.[15] Bowles himself provides some insight into the related themes of shelter and security in his autobiography, *Without Stopping*. The title of his novel, he revealed, came from a song popular in his childhood called "Down Among the Sheltering Palms." Bowles recalled, "It was not the banal melody which fascinated me, but the strange word 'sheltering.' What did the palm trees shelter people from, and how sure could they be of such protection?" (275).

TOURISM AND COLONIALISM

Bowles is also careful to take up the issue of European colonization in the non-Western world in several important ways. First, he introduces the Lyles, British subjects (and mother and son) who claim to have lived in both India and Africa for years, supporting themselves with the money generated from travel books authored by Mrs. Lyle. Though her books praise the picturesqueness of the various African countries they travel through, neither Mrs. Lyle nor her son, Eric, can tolerate any of the natives. Thus, the Lyles represent various strains of travel writing, as traced by Pratt in *Imperial Eyes*, that rhapsodize about exotic natural landscapes while representing the natives in pejorative terms (or ignoring them altogether).[16] The Lyle's close association with former British colonies, and the revelation later in the book that they have an incestuous relationship, positions them as the personification of a corrupt and decaying empire that has no regard for the natives of its African colonies. Though it is tempting to see the Lyles as counterpoints to Port and Kit, the fact that it is Eric who steals Port's passport suggests an identification between the two that would again undercut Port's notion that he is a traveler of a different sort from the Lyles.

But the setting of the novel is, of course, Algeria, still a French colony at the close of World War II, and Bowles is careful to examine French rule in the region. In perhaps the most interesting chapter in the novel, Bowles digresses from the story of Port and Kit at the beginning of Book Two to focus on Lieutenant d'Armagnac, commander of the French military post at Bou Noura. In this chapter, Bowles relates how the Lieutenant discovers a dead infant on his morning walk. Eventually, he tracks down the native mother, who confesses to the infanticide, and has her arrested. However, while awaiting trial in French custody, the mother is stung by a scorpion and dies. The aftermath of the young woman's death strikingly reveals the precarious and fragile nature of colonial power: the Lieutenant finds himself and his men ostracized by the native townsfolk, and soon after, the Lieutenant is poisoned by his native cook and almost dies.

The infanticide episode is designed by Bowles to facilitate a critique of European colonialism. For one thing, it is made clear that the Lieutenant and all his officers have sexual relations with native women. The lack of remorse on the mother's part, and the fact that the father of the dead infant is never sought by the Lieutenant, suggest that he was probably a French soldier. Like Port, the French soldiers view the native Algerian women as a kind of local resource to be used for their sexual pleasure. And through the Lieutenant's outrage at the infanticide, through his characterization of it as a savage barbarity, Bowles also creates an interesting counterpoint to the Lyle's incestuous relationship, a barbarity that disgusts but is tolerated and protected by the silence of the other Westerners whom they meet. Like the identity of the French soldier who fathers Yamina's child, the Lyle's relationship will remain hidden and protected.

It is here in Bou Noura that Port discovers that his passport has been stolen. Although he suspects Eric Lyle of the theft, Port goes to the French authorities and accuses Abeldekkar, the owner of the hotel where Port and Kit are staying. This accusation greatly upsets the Lieutenant, who, already on shaky ground with the native residents, realizes the important way in which Abeldekkar buttresses French authority in the region: "As proprietor of the inn he maintained close friendships with the chauffeurs of all the buses and trucks that passed through the territory. In the Sahara these are important people" (153). In other words, to a great extent, Abeldekkar controls the flow of goods and people in and out of Boa Noura. Thus, to secure and re-establish his own power, the Lieutenant must get Port to withdraw his accusation against him. Also consider the following scene, in which the Lieutenant orders a servant to bring more firewood:

> "More wood!" shouted the lieutenant, looking into the fireplace where the flames were dying down. But Ahmed refused to be prodigal with the wood, and brought in another small armful of the meager, gnarled branches. He remembered the early mornings of bitter cold when his mother and sister had got up long before dawn to set out across the high dunes toward Hassi Mokhtar; he remembered their return when the sun would be setting, and their faces, seamed with fatigue, as they came into the courtyard bent over double beneath their loads. The lieutenant would often throw on the fire as much wood as his sister had used to gather in the entire day, but he would not do it; he always brought in a scant amount. The lieutenant was quite aware that this was sheer recalcitrance on Ahmed's part. He considered it a senseless but unalterable eccentricity. (219)

Here again, Bowles conspicuously recounts an episode that highlights various aspects of the colonial situation. For the Lieutenant, the labor used to collect the firewood is never a consideration, or, at best, the people who labor to collect the firewood are as much an exploitable resource as the wood itself. For Ahmed, on the other hand, the Lieutenant's wastefulness reflects the French attitude toward his people, and his "recalcitrance," as the Lieutenant calls it, is really a form of native resistance that we could group with the poisoning of the Lieutenant and his ostracization by the town. Each of these examples, I think, could be used to support Papayanis's claim that expatriate fictions are not "indifferent to the relations of inequality that inevitably attend—that enable—the practices it depicts by looking at the problem of self and Other" (7). Thus, through his skillful presentation of this digression from the main narrative, Bowles illustrates the extreme native resentment toward the French rule imposed upon them, the fragile nature of colonial power, and the hypocrisy behind the civilized/primitive opposition upon which colonialism rests.

CONCLUDING REMARKS

Tucked away in the account of the infanticide, there is a tantalizing reference to a Frenchwoman who, many years prior, somehow ended up in the remote village of Bou Noura and gave birth to a young girl, fathered by an Arab: "What the Frenchwoman had been doing in Bou Noura all alone, no one ever knew. But it was all in the distant past" (152). Within the context of the narrative, this unnamed Frenchwoman has become nothing more than a dimly remembered legend, just as Kit is destined to be. But Bowles's tale, *The Sheltering Sky*, continues to live on, capturing the imagination of new generations of readers. It is a rich and complex novel, one that anticipates in astonishingly precise terms many of the concepts central to later theorizations of travel and migration. In it, Bowles does not simply portray a colonialist or expatriate mentality; rather, he provides a perceptive critique of the expatriates' colonizing relationship to the land and the people. In the process, he creates his own theory of why people travel and tour, his own theory of how travel and tourism often require and help to produce mythologized identities for both traveler and native, as well as his own theory of the economic exploitation that is rampant in these relationships and manipulated by everyone for their own agendas. And while Bowles appears, at first glance, to endorse the notion of a primitive Africa with the power to destabilize and destroy the Western mind, he also created a novel that, in the process of laying bare European colonialism, prompts his readers to ask themselves where the primitive really lies.

NOTES

1. The notion that tourism has replaced and made impossible a more meaningful type of travel is one that predates Fussell by many years. As Roger Bowen notes in his superb essay on Alex Garland's novel *The Beach*, Joseph Conrad makes a similar lament in several of his texts: "Conrad's anxiety about the multitudes that now have access to the globe, and about the erosion of the exotic and the obscure in a world so thoroughly mapped, is expressed vividly in his introduction to Richard Curle's *Into the East* and in his essay 'Geography and Some Explorers,' both composed at the end of his career" (43).

2. MacCannell defines the modern as post-industrial society: "Post-industrial or modern society is the coming to consciousness of industrial society, the result of

industrial society's turning in on itself, searching for its own strengths and weaknesses and elaborating itself internally. The growth of tourism is the central index of modernization so defined" (182).

3. See, for example, Andrew Martino, "The Vanishing Point: The Dis-Integration of Female Identity in Paul Bowles's *The Sheltering Sky*," *South Atlantic Review* 71.2 (2006): 87–114. A notable exception to the critical tendency to focus on dissolution of self in *The Sheltering Sky* is provided by Brian T. Edwards in his essay, "Sheltering Screens: Paul Bowles and Foreign Relations," *American Literary History* 17.2 (2005): 307–34.

4. Several critics have pointed out the overly simplistic and unconvincing nature of Fussell's distinction between travelers and tourists. In his essay "Semiotics of Tourism," Jonathan Culler suggests, "This repetition and displacement of the opposition between tourist and traveller suggests that these are not so much two historical categories as the terms of an opposition integral to tourism" (130). I would point out that tourists are just as often motivated by curiosity as travelers, while Fussell's travelers were just as often motivated by pleasure-seeking (as he himself acknowledges); and, as Susan Sontag, John Urry, and others have pointed out, tourism today is often a kind of work. Bowles, too, will go on to reveal the problematic nature of Port's assertion, as I discuss later in this chapter.

5. These ideas about the post-war flight from England to escape its limitations, as well as the horrors of mechanized warfare, are articulated by Fussell in the first three chapters of *Abroad* (3–23). The lure of the former colonial world is discussed throughout the book, but see, for example, Fussell's chapter, "The New Heliophily," (137–41).

6. Fussell writes, "One striking post-Second War phenomenon has been the transformation of numerous former small countries into pseudo-places or tourist commonwealths, whose function is simply to entice tourists and sell them things" (43).

7. Statements made by Bowles in at least one interview would suggest that he was sympathetic to such a characterization of the modern world. Speaking of *The Sheltering Sky*, Bowles commented that it was a story about trying to live in a world governed by indifferent forces, like nature: "Not caring. Unaware. And if you use the word God in place of nature, then I think you get even closer to it." Taken from an interview with Oliver Evans in the *Mediterranean Review* 1 (1971): 3–15, as quoted in Lawrence Stewart, *Paul Bowles: The Illumination of North Africa*, 53.

8. The idea that too often we take life for granted is one of the novel's major themes and recurs throughout. It is the basis for one of the novel's most famous and beautifully-written passages. "And Port had said: 'Death is always on the way, but the fact that you don't know when it will arrive seems to take away from the finiteness of life. It's that terrible precision that we hate so much. But because we don't know, we get to think of life as an inexhaustible well. Yet everything happens only a certain number of times, and a very small number, really. How many more times will you remember a certain afternoon of your childhood, some afternoon that's so deeply a part of your being that you can't even conceive of your life without it? Perhaps four of five times more. Perhaps not even that. How many more times will you watch the full moon rise? Perhaps twenty. And yet it all seems limitless'" (238).

9. Patteson astutely notes that Port's foray from his hotel to the Arab encampment on the outskirts of the city is the first of several journeys in miniature that foreshadow the later movements in which return becomes increasingly difficult and, ultimately, impossible—as succinctly expressed in the telegram Kit wants to send, "CANNOT GET BACK." See Patteson, 98.

10. Pratt, 6.

11. Allen Hibbard does an especially good job of locating Bowles within the Orientalist tradition as it pertains specifically to North Africa. See his book, *Paul Bowles: Magic and Morocco*, 20–34.

12. Patteson, 56; Hibbard, *Magic and Morocco*, 20–34.

13. For the original interview, see Jeffrey Bailey, "The Art of Fiction LXVII: Paul Bowles," *Paris Review* 81 (1981): 63–98.

14. Hibbard notes that Bowles dedicated *The Delicate Prey and Other Stories* (in which "A Distant Episode" appears) to his mother, "who first read me the stories of Poe." See Hibbard's *Paul Bowles: A Study of the Short Fiction*, 12. Bowles himself acknowledged "A Distant Episode," written in 1945, as an inspiration for *The Sheltering Sky*. See Patteson's introduction, x.

15. I am grateful to my student, Rachel Egboro, for this insight.

16. See Pratt's chapter, "Narrating the Anti-Conquest," (38–68).

Chapter Two

The Ruins of Manifest Destiny

John L. Stephens (1805–1852) studied law at Columbia University and then moved into political campaigning, but illness interrupted this work and became a turning point in his life, as doctors advised him to travel abroad to aid his recovery. The books resulting from those travels, *Incidents of Travel in Egypt, Arabia Petraea, and the Holy Land* (1837) and *Incidents of Travel in Greece, Turkey, Russia, and Poland* (1838), sold well, and Stephens was able to use some of the profits to finance a trip to Central America, which he hoped would form the basis for another travel book.[1] The selection of Central America as a destination was not arbitrary. Stephens's collaborator, the illustrator Frederick Catherwood, had introduced him to an account of mysterious, ruined cities in Mexico and Central America, and the two men began to plan a trip to the region.[2] The journey then became possible when US President Martin Van Buren appointed Stephens as US *charge d'affaires* to Central America and gave him the task of ascertaining whether a viable government existed with which diplomatic relations could be established. In 1839, Stephens and Catherwood set out for Central America, a country in the midst of civil war.

Reports of the deserted Maya cities had begun to surface in the United States and Europe only in the decade following Mexican independence from Spain in 1821. During the Spanish colonial period, Spanish officials closely guarded information about the Spanish American colonies. As R. Tripp Evans explains:

> Prior to the expulsion of Spain from Latin America in the 1820s, the very existence of most pre-Columbian sites was generally unknown. Their rediscovery following Spain's departure initiated a period of exploration that the New World had not seen since the sixteenth century, undertaken at a time when ancient geography, biblical history and even the source of human origins itself were all being seriously questioned. The ruins of Mexico and Central America

in the era were for a variety of different intents and purposes a historical tabula rasa. (Evans 2)

Stephens's account of his Central American journey, *Incidents of Travel in Central America, Chiapas, and Yucatán*, first published in 1841, recounts his visits to such now-famous Maya sites as Copán, Quiriguá, Palenque, and Uxmal. Unlike the work of his predecessors, Stephens's Maya books were relatively inexpensive, conversational in style, and contained Catherwood's extraordinarily precise illustrations of the ruins, stelae, and monuments. The books became bestsellers. According to Evans, "more than twenty thousand copies of their two-volume set sold within the first three months of printing" (70). Thus, it was largely Stephens and Catherwood who introduced the Maya world to an English-language readership. But Stephens's account is also notable as a historical document that reflects the growing expansionist designs of the United States, as well as the related, ongoing projects of nationalism and nation-building. One of the ways his text functioned in this regard, as Bruce Harvey points out, was by appropriating an indigenous American history and conjoining it with US history to fortify the nascent US republic's identity (158). Evans concurs: "Stephens' cultural enforcement of the newly established Monroe Doctrine, combined with his interpretation of the United States' 'manifest destiny' on the North American continent, constituted a double-barreled campaign to claim the Mesoamerican past as the United States' cultural inheritance" (44).

This campaign to claim the Mesoamerican past for the United States is evident in Stephens's American travel books (he and Catherwood would publish a sequel two years later in 1843, documenting their exploration of Maya sites in Yucatán) in both a figurative and a literal sense. Stephens tried to purchase, usually unsuccessfully, almost all of the Maya sites he visited on his first journey. In the most notorious episode, he purchased Copán for fifty dollars. Ever the entrepreneur, Stephens hoped to eventually open a museum housing Maya antiquities in New York City as a profitable enterprise. But, as Evans astutely suggests, Stephens's ambitions in this regard reflect the acquisitive ideology of an entire nation: "Stephens's compulsion to own the sites he explored not only indicated personal ambition but also reflected the United States' dramatic territorial expansion of the early nineteenth century" (55).

Although the term Manifest Destiny was not coined until 1845 by magazine editor John O'Sullivan, the ideological principles that it advocated were powerful forces in American society for at least four decades prior to the invention of the term, according to Harvey. Those principles, Harvey explains, grew out of "Assumptions about the superiority of the Anglo-Saxon race and republican government, about the legal right of advanced nations to appropriate undeveloped land, about the Protestant duty to spread the Gospel to heathen realms: all helped to justify the acquiring [sic] of new continental dominions" (7). Although the term Manifest Destiny is most often used in conjunction with US territorial expansion, I want to extend its use to a second phase of US imperialism: the economic expansion into Latin America that begins following the end of the war between the United States and Mexico in 1848 and which would be justified under the principles set out in the Monroe Doctrine of 1823. Although there would continue to be new territorial acquisitions by the United States, most notably following the conclusion of the Spanish-American war in 1898, in many ways, the signing of the Treaty of Guadalupe Hidalgo, formally ending the war between the United States and Mexico in 1848 and adding an enormous new geographic space to the United States, signaled a new phase of US imperialism, that of capitalist acquisition, and it is just such economic ambition that forms an integral part of Stephens's travel narrative, helping lay the groundwork for later capitalist intervention in Latin America. In this chapter, I hope to show how Stephens narrates "the incidents" of his travels in such a way as to further the idea of Mexico and Central America as potential areas for the United States to control and exploit for economic benefit, and to demonstrate that the popularity of his travel books with the American reading public, at least in part stemmed from the way they confirmed American cultural superiority and Americans' right to intervene in the hemisphere. As William Lenz puts it, "As [Stephens] writes his narrative, and as we read his narrative, Americans possess Central America and absorb its history" ("Identity" 87). Lenz observes that Stephens presents himself as the embodiment of an idealized American character, in a text that vividly reflects the cultural anxieties of the period and attempts to allay those anxieties by repeatedly contrasting a decadent, corrupt, and violent Central American society against a civilized and secure democratic United States, even as the latter wrestles with such issues as slavery and the forced removal of Native Americans from their lands. Lenz argues that, taken in its entirety, *Incidents* reinforces the superiority of American values, which it defines as secular, industrious, and progressive. He sums up the tremendous appeal of *Incidents* for its nineteenth-century American audience this way:

> To focus attention in Central America, away from the national debates about States' Rights, slavery and Free Soil, Nat Turner's rebellion, the Indian Relocation Acts, the economy in the wake of the Panic of 1837 and the suspension of specie payments by regional banks, consolidated popular public confidence in a shared vision of national identity. Americans–whether in Boston or Richmond–reading Stephens' narrative could share in his successful appropriation of the Central American landscape and Mayan past to an imagined, progressive, optimistic, Franklinian, unified, national future. That it might be in fact illusory is irrelevant to Stephens' literary success. Stephens' narrative

provides a distraction from contemporary social problems. In this view, it confirms deeply felt cultural values. It is a reaffirmation of the myth of national success. (*Ruins* 67)

As Lenz perceptively notes, Stephens very quickly comes "to feel himself as the embodiment of American nationalism" (81). For example, Stephens asserts the racial superiority of whites even in the face of contradictory evidence (e.g., the black communities he encounters in Belize, the racially mixed communities he encounters in Central America and Mexico, etc.), and he literally cloaks himself in the symbolic protection of the American eagle in moments of dire peril, even as he plots to appropriate Maya artifacts and dreams of constructing a transcontinental canal across the Central American isthmus (Stephens's death in 1852 was the result of tropical illnesses he contracted while overseeing the construction of a railroad across the Isthmus of Panama).

As a member of the capitalist vanguard, to use Mary Louise Pratt's term, Stephens imagines numerous ways that the United States government (and he himself) might profitably exploit these local Latin American resources, even as his narrative depicts him as clearly ill-equipped to understand or appreciate the cultural differences between the United States and Latin America.[3] Stephens's descriptions of his stay in Belize, en route to Guatemala, provides a useful starting point. Here he encountered a predominantly black community governed by a white English minority that had the potential to challenge his ideas about race. Unlike the United States, Belize had legally abolished slavery, and it appeared to Stephens that the African Americans there were relatively well-integrated into the social structure and had a certain degree of social mobility. In the following passage, Stephens describes being seated between two black gentlemen at a dinner in his honor at the colonial government house:

> Some of my countrymen, perhaps would have hesitated about taking it, but I did not; both were well dressed, well educated, and polite. They talked of their mahogany works, of England, hunting, horses, ladies, and wine; and before I had been an hour in Balize, I learned that the great work of practical amalgamation, the subject of so much angry controversy at home, had been going on quietly for generations; that colour was considered mere matter of taste; and that some of the most respectable inhabitants had black wives and mongrel children, whom they educated with as much care and made money for with as much zeal, as if their skins were perfectly white. I hardly knew whether to be shocked or amused at this condition of society. (*Central America* Vol. I, 12–13)

There are at least two points of interest here: With the slavery debate tearing up the United States at this historical moment, Stephens points to Belize as an example of what an integrated society without slavery might look like. But although Stephens begins by emphasizing his open-mindedness, his remarks reveal that he is anything but and that he finds this racially-integrated society both shocking and amusing. Although "colour" may be considered a "mere matter of taste" in Belize, it certainly is not for Stephens, nor to the British colonists, as his subsequent observations make clear. African Americans may be more integrated into this colonial society, but it soon becomes clear that they comprise the working class of the colony and are relied upon for any work involving manual labor. And while slavery has been legally abolished, a form of indentured servitude is permitted. Moreover, miscegenation may be common and socially acceptable, but the racially-mixed society it produces is very carefully ordered and regulated by a caste system defined by the English, based on racist ideas of white superiority. As one schoolteacher tells Stephens, "The brightest boys, and those who had improved the most, were those who had in them the most white blood. . . . white scholars were always the most quick and capable" (Vol. I, 16–17). Stephens reports such remarks without raising any objection to them, and a thorough reading of his travel books suggests that he shares the schoolteacher's views of white supremacy. For example, he remarks with surprise upon how completely civilized an isolated Carib community has become without mingling their blood with that of their conquerors and goes on to represent certain of their members atavistically, in particular one old woman whom he tells the reader "looked as though in her youth she had gloried in dancing at a feast of human flesh" (Vol. I, 29). As Lenz puts it, "Stephens thrills at his own proximity to an excitingly savage cannibal. Even if she is not a real cannibal, her 'idiotic laugh' insures that we [the readers], like Stephens, associate her with forces of barbarism" (82). With regard to the Latin American women he meets during his travels, Stephens consistently remarks upon their physical features. Although cloaked in innuendo, he never tires of remarking on the number of native women he encounters who bare their breasts.

Rhetorically, Stephens's narrative is a virtual compendium of the different modes of travel writing outlined by Pratt in her important study, *Imperial Eyes*. During the description of Maya sites, Stephens slips into the scientific, informational mode—describing, reporting, measuring, cataloging. At other points during his travels, he has romantic reveries as he gazes upon untamed wild nature. But by far the two rhetorical modes that predominate are what Pratt calls the capitalist vanguard and the sentimental, which, to a certain extent, paradoxically complement and work against each other. The "capitalist vanguard" are those travelers who evaluate other nations and cultures predominantly in terms of economic development, usually focusing on their alleged lack of development and the tremendous potential for profit if the countries' resources were turned over to North American or European capitalists. Thus, we find Stephens frequently berating the lack of infrastructure in Central America and continually on the lookout for economic opportunities. What follows are three representative passages. In the first, Stephens makes a case for North American construction and control of a canal across the Central American isthmus:

> It will compose the distracted country of Central America; turn the sword, which is now drenching it with blood into a pruning hook; remove the prejudices of the inhabitants by bringing them into close connection with people of every nation; furnish them with a motive and a reward for industry, and inspire them with a taste for making money, which, after all, opprobrious as it is sometimes considered, does more to civilize and keep the world at peace than any other influence whatever. (Vol. I, 419)

As Miguel Cabañas observes, "Stephens portrays the problem within the dichotomy of progress and anarchy. War prevents those countries from reaching a utopian future of progress and prosperity and Stephens believes that a capitalist influence would deliver peace" (50).

My second example concerns Stephens's plans to purchase, remove, and ship the Maya antiquities at Copán to the United States to be housed in a museum—a business enterprise he imagines will be very profitable. Stephens justifies his scheme on the grounds that:

> Other ruins might be discovered even more interesting and more accessible. Very soon their existence would become known and their value appreciated, and the friends of science and the arts in Europe would get possession of them. They belonged of right to us, and, though we did not know how soon we might be kicked out ourselves, I resolved that ours they should be; with visions of glory and indistinct fancies of receiving the thanks of the corporation flitting before my eyes, I drew my blanket around me, and fell asleep. (Vol. I, 115–16)[4]

This oft-quoted passage, as many critics have remarked, is a striking example of Stephens's invocation of a kind of cultural Monroe doctrine.[5] Nor, did Stephens's plans stop with transporting individual artifacts to be housed in a museum—he also entertained notions of trying to transport entire Maya cities to the United States in order fence them off and charge admission to curious tourists. Although such an undertaking was impractical in his day, Stephens's vision of a prototypical Mayan theme park now seems prescient, given how carefully orchestrated and organized the sites are today, as James Clifford details in his essay on the Palenque ruins.[6]

Even volcanos bring out the capitalist in Stephens. My final and favorite example occurs as Stephens gazes into the crater of a volcano after a long climb in Nicaragua: "I could not but reflect, what a waste of the bounties of

Providence in this favoured but miserable land! At home this volcano would be a fortune; with a good hotel on top, a railing round to keep children from falling in, a zigzag staircase down the sides, and a glass of iced lemonade at the bottom" (Vol. II, 13).

Stephens's official reason for traveling to Central America, we should keep in mind, was to ascertain whether a viable government existed with which the United States could establish diplomatic relations. This diplomatic mission is presented in terms that further the notion of the region as one where US intervention would be warranted, and Stephens concludes that there is, in fact, no viable government to work with. Noting that Stephens uses archaeological terms in describing the political situation, Evans remarks:

> Whereas his search for a government proved fruitless, he was successful in locating the ruins; whereas the government was "in ruins" and "beyond restoration," he hoped for the salvation of the areas's antiquities. Ultimately, in his conclusion that regional government had ceased to exist, he implicitly declared the archaeological landscape to be politically unaffiliated. (62)

And thus, of course, it is easier to claim for the United States.

All this desire for economic exploitation finds itself undercut in the narrative by the sentimental rhetorical mode. As Pratt explains, in the sentimental travel narrative, the narrator dramatizes himself or herself as protagonist, one who is at the mercy of weather, disease, cultural misunderstandings, and misadventures in ways that leave the narrative open to question the capability of such outsiders to intervene in the affairs of the host countries. As is typical of these narratives, there is a tension between what the traveler claims about the indigenous population (e.g., they are indolent, lazy, selfish, ignorant, etc.) and how the travels are carried out. Stephens is completely dependent upon the Indians whom he denigrates for guidance, transportation, and the portage of his heavy equipment over difficult terrain. And because Stephens is traveling through remote areas that rarely have travelers passing through, he must rely upon the goodwill of the people he meets along the way. Failing this, he offers to pay for food and lodging (since his diplomatic passport carries little weight with the common folk) and goes to such extremes as paying poor families to sleep outside so that he can have the house to himself for the night. In fact, one of the things that infuriates him most is that so many of the people he encounters are unwilling to be bought off in this manner. Likewise, he is frustrated by potential guides who refuse to take on dangerous journeys, even at the high prices he offers to pay. And, as David Johnson points out, while he relies on native informants and guides to lead him to various Maya ruins, he claims that they are completely ignorant about the ruins' significance—an assertion which, of course, Stephens uses to strengthen his argument for ownership of the Maya sites (8–9).

One of the most fascinating episodes in Stephens's narrative has to do with just such a moment of the foreign traveler's dependence upon native labor. During a particularly arduous stretch of mountain terrain en route to Palenque, Stephens fell ill and was badly weakened by fever. Physically incapable of ascending the rough, mountainous terrain, he was forced to be carried by a native *sillero*, who carried Stephens on a chair strapped to his back. The subsequent loss of control and the feeling of being physically bound to a native Other immediately unsettle Stephens. He describes his reactions in the following passage:

> [The *silla*]was a large clumsy armchair, put together with wooden pins and bark strings. The Indian who was to carry me, like all the others, was small, not more than five feet seven, very thin, but symmetrically formed. A bark strap was tied to the arms of the chair, and, sitting down, he placed his back against the back of the chair, adjusted the lengths of the strings, and smoothed the bark across his forehead with a little cushion to relieve the pressure. An Indian on each side lifted it up, and the carrier rose on his feet, stood still a moment, threw me up once or twice to adjust me on his shoulders, and set off with one man on each side. It was a great relief, but I could feel every movement, even to the heaving of his chest. The ascent was one of the steepest on the whole road. In a few minutes he stopped and sent forth a sound, usual with Indian carriers, between a whistle and a blow, always painful to my ears, but which I never felt so disagreeably before. My face was turned backward; I could not see where he was going, but observed that the Indian on the left fell back. Not to increase the labour of carrying me, I sat as still as possible; but in a few minutes, looking over my shoulder, saw that we were approaching the edge of a precipice more than a thousand feet deep. Here I became very anxious to dismount; but I could not speak intelligibly, and the Indians could or would not understand my signs. . . . I rose and fell with every breath, felt his body trembling under me, and his knees seemed giving way. The precipice was awful, and the slightest irregular movement on my part might bring us both down together. I would have given him a release in full for the rest of the journey to be off his back; but he started again, and with the same care ascended several steps, so close to the edge that even on the back of a mule it would have been very uncomfortable. My fear lest he should break down or stumble was excessive. To my extreme relief, the path turned away; but I had hardly congratulated myself upon my escape before he descended a few steps. This was much worse than ascending; if he fell, nothing could keep me from going over his head; but I remained till he put me down of his own accord. The poor fellow was wet with perspiration, and trembled in every limb. Another stood ready to take me up, but I had had enough. . . . It was bad enough to see an Indian toiling with a dead weight on his back; but to feel him trembling under one's own body, hear his hard breathing, see the sweat

> rolling down him, and feel the insecurity of the position, made this a mode of traveling which nothing but constitutional laziness and insensibility could endure. (Vol. II, 274–76)

This passage is remarkable in several ways. First and foremost, the colonial dependence upon indigenous labor is physically embodied in the scene. But even more fascinating is the fusion of Stephens with what is, for him, a native Other in a way that shifts control to the Indian carrier. This loss of control clearly unsettles him, but even more disturbing is the sensation that he suddenly occupies the same *body* as the native; Stephens senses every movement and even every breath that the Indian makes. Finally, the scene destroys the myth of the omniscient, self-sufficient foreign traveler. Here, Stephens is weak, powerless, unable to see where he is going, and completely dependent upon native support to succeed in his journey, and it is the European/Anglo, not the native, who is represented as unable to speak intelligibly.

The fame of Stephens's books rests on the Maya sections, and with good reason. Catherwood's painstaking illustrations of stelae, statues, and temples are so accurate that, even today, Mayanists refer to them to read glyphs that have eroded beyond the point of recognition, and Stephens's theories as to why the Maya cities fell into decline—war, famine, or pestilence—continue to be those most widely held by contemporary Mayanists. He also astutely recognized the Maya hieroglyphs as a form of writing in which the history of the Maya is, in his words, "graven on its monument" (Vol. I, 159–60). Having recognized the glyphs as a form of writing, Stephens was clearly frustrated at not being able to decipher them. Thus, as both Harvey and Johnson note, the Maya ruins that Stephens visited challenged him on both a literal and a figurative level. Because so many of the Maya buildings and structures Stephens encountered were filled with debris or overgrown with vegetation, he was physically unable to penetrate to a hidden Maya interior, just as the Otherness of the Mayan language and its written form figuratively prevented him from penetrating to an interior meaning. Harvey suggests that Stephens resolved this frustration through his narration of events by transforming the archeological failures into a successful literary operation that involved gazing upon Central American vistas while imagining pre-Columbian or Spanish colonial scenes of Stephens's fantasies (Harvey 170–71). What cannot be accomplished in reality is accomplished in the realm of the imagination. In perhaps the most remarkable example of this tendency, Stephens imagines leading a military force to conquer a still-populated Maya city rumored to exist deep in the jungles of Chiapas. Evans comments: "Stephens essentially proposed a re-enactment of the Spanish conquest. By seeking cultural treasures, rather than gold, however, he justified this invasion on scientific grounds" (69).

In addition to such anachronistic passages, we might consider Stephens's narrative to be a travel fiction based on his use of other novelistic devices. As noted earlier, his travel books were very popular in the nineteenth century and continue to be today.[7] This is not, I think, because Stephens is a particularly good writer. His narratives have, as Harvey aptly puts it, a "compulsive seriality" that can make them quite tedious for long sections (107). However, Stephens did have a wry sense of humor, and he frequently poked fun at himself over cultural misunderstandings, misadventures, and being ill-prepared to deal with certain unexpected circumstances. These self-deprecatory passages go a long way toward earning his readers' sympathy. On other occasions, though, that sympathy for himself is developed at the expense of the natives he meets. Early in the narrative, this role is played by the guides, porters, and muleteers who are often portrayed as incompetent, lazy, and stubborn. The expedition's young guide, Augustin, for example, stores a packet of gunpowder together with the food supplies, and during the overland journey, the packet comes apart, spoiling the food and making it inedible. But Stephens's greatest nemesis, by far, in the early part of the narrative is an unnamed "rascally muleteer" who is portrayed as increasingly recalcitrant as the journey to Copán continues, due to a falling out with Stephens overpayment for services. While Stephens portrays the disagreement as one in which the muleteer is simply trying to bilk the American travelers for more money than was originally agreed to, we learn later that, in fact, there was a misunderstanding between the two parties over when payment was to be made: Stephens insisting that there will be no payment until arrival at Copán, and the muleteer demanding payment be made at regular intervals en route. This is a good example of how Stephen manipulates the narrative in his favor: the muleteer is never given a name, allowing Stephens to more easily represent him as a villain of sorts, and as narrator of the tale, Stephens is able to represent the muleteer as stubborn and recalcitrant, obscuring the more likely possibility that Stephens and Catherwood were making demands that were unwise, difficult to carry out, or even impossible. That their falling out was about money is also revealing, as it sheds light on why the muleteer and the other porters and guides would have been willing to accede to demands they thought unwise or even dangerous. We see the antagonism that such economic disparity engenders even today in relations between foreign tourists and their hosts. The other villain of the Copán episodes is Don Gregorio, the *hacendado* who governs the area around the ruins. Although Stephens resents his lack of hospitality and suspicion of the foreigners, those suspicions turn out to be well-founded as Stephens and Catherwood hire his workers away to help with their excavations of the ruins, even as they scheme to remove and ship the Mayan antiquities back to the United States.

The Copán chapters also highlight the ways that Stephens borrows from and reworks the conventions of earlier types of travel narratives, in particular, the captivity narrative and the narrative of exploration. For example, Stephens recounts episodes in which he and Catherwood are asked by local inhabitants to remedy various medical ailments, and their attempts to practice medicine are cited by Stephens as one of the reasons they are permitted continued access to the ruins, as well as a cumbersome responsibility:

> We had so many applications for *remedios*, our list of patients had increased so rapidly, and we had been so much annoyed every evening with weighing and measuring medicines, that, influenced also by the apprehensions before referred to, we had given out our intention to discontinue practice; but our fame had extended so far that these people had actually come from beyond San Antonio more than thirty miles distant to be cured, and it was hard to send them away without doing something for them. (Vol I., 147–48)

Such episodes were a staple of narratives of exploration of the sixteenth and seventeenth centuries, as well as ones recounting the use of unfamiliar technology as a tool of conquest, reworked by Stephens in scenes that parody them: "In the morning we continued to astonish the people by our strange ways, particularly by brushing our teeth, an operation which, probably, they saw then for the first time" (Vol. I, 106).

Also notable is Stephens's adaptation of one of the key features of the captivity narrative: the idea that the captive survives through his or her religious faith. In Stephens's account, survival is attributed to patriotic, not religious, devotion. This revision of the genre is most apparent in Stephens's recounting of an episode en route to Copán. Arrested by town officials in Comotán, Stephens presents them with his diplomatic passport, which they refuse to recognize as legitimate. An argument ensues, and Stephens finally persuades his captors to send a letter by courier to other officials in a nearby town to verify his diplomatic status; however, he is stymied by the lack of an official signet to seal the letter. In an inspired moment of improvisation, he devises a solution: "We sealed it, unobserved by anybody, with a new American half dollar, and gave it to the alcalde. The eagle spread his wings, and the stars glittered in the torchlight" (Vol. I, 82–83). A few hours later, the travelers are released and allowed to continue their journey. Stephens wonders over this rapid turn of fortune and concludes, "I have no doubt that the big seal did much in our behalf" (84). Here, the captive turns not to God for deliverance but instead to the nation, and Stephens cloaks himself in the symbolic protection of the American eagle, attributing his release to the power of that nationalist symbol, a motif that recurs throughout the narrative. In the notorious episode in which he purchases Copán for fifty dollars, Stephens claims that the seller agreed to part with the property because he "could not withstand the buttons on my coat" decorated with eagles (Vol. I, 128). Stephens is so impressed with these results that he affixes an eagle insignia to the brim of his hat. Such episodes highlight how closely wedded the nationalist symbols of the United States are with the notion that it is the United States' manifest destiny, bestowed by Divine Providence, to overspread the continent, at this historical moment. The use of a piece of US currency as a diplomatic seal says volumes about the international politics of the United States of that era, and the eagle, that ubiquitous symbol of American imperialism, hovers over the remainder of the narrative like a watchful predator. As one of the most popular American books of the mid-nineteenth century, Stephens's *Incidents of Travel in Central America, Chiapas, and Yucatán* clearly reinforces that era's widespread American belief in the United States' right to intervene in Latin America—and to a large extent such reinforcement is accomplished through Stephens's skillful construction of a travel fiction about Latin America. David E. Johnson's apt observations regarding travel narratives set in Latin America perfectly sum up Stephens's Maya travel books: "American travel writing traces the labyrinth of its itinerary on the surfaces of the Americas, grounding the politics of the Americas in the fictions of travel: Anglo-North American linguistic superiority, 'indigenous' ignorance, the privilege of European aesthetics over Amerindian cultural productions, the emptiness of the Latin Americas, and the rule of the alphabet" (24).

CONCLUSION: INTERDEPENDENT GENRES

In closing, I would like to return to one of the observations I underscored in my introduction, specifically the idea that travel narratives and travel fictions are extremely interdependent genres, so much so that they are two sides of the same coin. Stephens's text and its reception perfectly illustrate this relationship. When *Incidents of Travel in Central America, Chiapas, and Yucatán* was published in 1841, one of its most avid readers was the historian W. H. Prescott, who gave a copy to Frances Calderón de la Barca, wife of the Spanish ambassador to Mexico. During her time in Mexico, Calderón corresponded with Prescott and, at his urging, published her letters in revised form as a travel narrative, *Life in Mexico*, in 1842. As I have detailed elsewhere, Calderón's travel narrative provides an interesting, gendered counterpart to that of Stephens. Instead of focusing on volcanos and pyramids, Calderón presents a world of high society—soirees, costumed balls, and bullfights. Both of their travel narratives became important sources for Prescott and greatly influenced him as he was writing his massive history of the conquest of Mexico, published in 1843, which, for many years, was the only available history of the conquest in the English language. Prescott's literary bent produced a history narrated as epic drama, its actors superhuman figures. The impact of Prescott's history and the effect of his tendency to romanticize the Conquest as heroic drama should not underestimated. The "Prescottian fable," as Inga Clendinnen calls it, taught its readers that "Europeans will triumph over natives, however formidable the apparent odds, because of cultural superiority, manifesting itself visibly in equipment but residing much more powerfully in mental and moral qualities"—qualities, she argues, that Prescott coded into his oppositional renderings of Cortés and Montezuma: the "ruthless, pragmatic, single-minded" Spanish commander juxtaposed against the "despotic, effete, and fatally indecisive" Aztec emperor (65–66). Clendinnen's point, of course, is that, in Prescott's history, Cortés and Montezuma function as cultural synecdoches, a representation that, unfortunately, has been reinscribed by many later historians (most notably Tzvetan Todorov). And historians were not the only ones who found the Prescottian fable irresistible. In the foreword to his 1893 novel *Montezuma's Daughter*, H. Rider Haggard acknowledges drawing freely from Prescott for details of his own very romantic version of the Conquest, in which a dashing English hero arrives in Mexico just in time to educate Doña Marina and aid "Cuactemoc" in his battle against Cortés. This chain of literary influence continues to the present day. Graham Greene, who wrote both Mexican travel narratives and travel fictions, attributed his interest in Mexican culture to Haggard's novel. The sense of these travel narratives and travel fictions operating in an interdependent and systematic fashion, as ideas of cultural superiority and inferiority are borrowed, reinscribed, and passed on by later authors, begins to impress upon us its significance.

NOTES

1. According to Bruce Harvey, sales of Stephens's first two travel books earned him $25,000 and helped to bankroll his travels in Central America (105).

2. Harvey provides this account: "When in London in 1836, about to return home from his trip to the Levant, Stephens had befriended Frederick Catherwood, the English artist then gaining fame for his exhibit of a vast panorama of Jerusalem. Catherwood showed Stephens a book by Antonio del Río, a captain in the Spanish army, who in 1786 had stumbled upon the ruins of Palenque in Southern Mexico. The Mayan cities, buried by the dense debris of the Mesoamerican rain forest, had before Rio's account remained the subject of only vague rumor. The soldier's discoveries excited Stephens, and once back in New York he came across several recently published studies that further piqued his interest, especially Jean Frederic Waldeck's lavish folio volume *Voyage Pittoresque et Archeologique dans la Province d'Yucatán* (1838)." (Harvey 161).

3. See Mary Louise Pratt, *Imperial Eyes: Travel Writing and Transculturation*, Routledge 1992.

4. In a terrible irony, many of the priceless and irreplaceable Maya artifacts collected by Stephens over the course of his travels were completely destroyed in 1842 when the New York City building housing them burned to the ground. For a discussion of this tragic event, see Evans, 73, and Leask, 140.

5. Miguel Cabañas makes this point in his insightful discussion of Stephens in his book on nineteenth-century travel narratives on 49–50.

6. See Stephens, *Incidents of Travel in Yucatán*, Vol. I, 136, for a representative example of Stephens's plans for a Maya theme park, and Clifford's "Palenque Log," in *Routes: Travel and Translation in the Late Twentieth Century*, 220–37.

7. Along with authors William H. Prescott and Frances Calderón de la Barca, Stephens was part of a trio of American writers whose books greatly shaped public perceptions of Mexico and the former Spanish American colonies. See Anna Brickhouse's *Transamerican Literary Relations and the Nineteenth-Century Public Sphere* for a discussion of Stephens's literary influence upon other authors.

Chapter Three

Travel Writing and the Tramp

From almost its inception, American literature has been notable for the wanderlust threading its way through many of its most famous works, often accompanied by an ambivalence regarding society and its rules and regulations. Think of Rip Van Winkle's desire to ramble through the woods rather than do his daily chores or Huck Finn chafing under the moral instruction of the Widow Douglas (Etulain 26). This archetypal theme emerged again in another incarnation, during the last quarter of the nineteenth century, in the figure of the hobo or tramp. This time, however, the flight from society was firmly rooted in the economic and technological changes transforming the United States in the post-Civil War period. As many historians have observed, the American tramp or hobo appears to emerge around the time of the economic depression of 1873, which left many Americans unemployed and/or homeless,[1] and the increasingly well-developed railroad system that crisscrossed the country provided a means for these unemployed migrants to travel in search of better circumstances, or merely to survive.[2] In addition, as Todd DePastino notes, large, corporate-owned farming, mining, and logging operations in the western United States created a demand for migrant labor on a seasonal basis (*The Road* xxii). All these factors contributed to the rapid increase in the numbers of traveling unemployed and homeless men and women during the last three decades of the nineteenth century. The increase was so sudden and dramatic that it brought about a new meaning for the term tramp, which, prior to 1873, had never been used to refer to an itinerant homeless or jobless person. By 1875, the term was firmly entrenched in American public discourse, and the issue of what to do about the "tramp crisis" was featured regularly in magazine articles and newspaper editorials.[3]

The coining of this term to refer to a homeless traveler or poor, itinerant worker suggests that there was something new or novel about the tramp, and in many regards, that was the case; however, there have always been traveling poor. Tim Cresswell notes that in Europe, the homeless traveler without means, most often referred to in England as a vagabond or vagrant, was historically perceived and treated as a threat to social order—and one that must be regulated—since at least the Middle Ages (17). In both early modern Europe and colonial America, vagrants were viewed with suspicion and were criminalized, subject to punishments that included flogging, branding, and execution: "Indeed vagrancy laws had been on the books of most [American] states since their foundation. For the most part, these had been borrowed wholesale from Great Britain, which had a long history of anti-vagrancy laws. The legal concept of vagabondage originated, in England, in the long-standing distinction between able-bodied and non-able-bodied poor" (50). DePastino concurs: "Like the laws of settlement, vagrancy statutes legitimized and facilitated the mobility of better-off transients while discouraging and criminalizing the movement of the poor" (*Citizen Hobo* 6).

The policing of vagrants has always centered on the absence of two things: a fixed home and demonstrable work skills or occupation. Skilled migrant workers have always fared better than unskilled ones, in both Europe and colonial and antebellum America. Indeed, as Judith Adler observes, in Europe, from the seventeenth through the mid-nineteenth century, working tramps were perfectly respectable members of society, usually skilled artisans who traveled from town to town, providing important services to the communities on their routes and supported by a network of guilds, unions, specified travel routes, and formal hospitality obligations developed to support these traveling workers. Adler notes that an important shift, due to economic factors, occurred during the 1870s in Europe as well as in America, when the "skilled artisan of the nineteenth century was replaced on the road by unskilled migrant workers, relying upon generalized public charity rather than craft hospitality, and finding accommodation in workhouses, spikes, charity missions, or hobo jungles rather than craft-affiliated inns" (341). Thus, from 1870 until the end of the Great Depression, the tramp increasingly came to be viewed not as an important member of society but rather as a dire social problem and even threat, and young people who became tramps were seen as juvenile delinquents: "The homeless man would recur as a subject of public debate and inquiry for the remainder of the [nineteenth century], with a new peak of interest in the mid-1890s. By that time, the tramp had become a powerful symbol both of crime and vice, including potentially 'perverse' sexual behavior" (Nissen 75–76). The increasingly well-developed railroad network in the United States was often cited as contributing to this problem. In Josiah Flynt's *Tramping with Tramps: Studies and Sketches of Vagabond Life* (1899), one of the most widely read and influential books on the subject of that era, Flynt remarks:

> A child possessed of wanderlust generally wanders for a while anyhow, but the chance he now has to jump on a freight train and "get into the world quick," as I

> have heard lads of this temperament remark, has a great deal to do with tempting him to run away from home. Hoboland is overrun with youngsters who have got there on the railroads. . . . The railroad, however, makes it possible for them to keep shifting the scenes they enjoy, and in time change and variety become so essential that they are unable to settle down anywhere. They are victims of what tramps call the "railroad fever." (Flynt 113, as quoted in Adler 343)

One of these young tramps was Jack London. In 1892, when he was just sixteen years old, London swung aboard a train in Sacramento and rode it to Reno. Two years later, he began a six-month journey, much of it by hopping on trains and tramping, that would take him from Oakland to Hannibal, Missouri. Along the way, he joined Charles T. Kelly's Industrial Army, a large group of unemployed men who had commandeered a freight train and were traveling from California to Washington, DC, to protest the lack of jobs created by the economic crisis of 1893. Eventually parting ways with Kelly's group, London spent time with an aunt in Michigan, then traveled by train (this time as a passenger) to New York City. His tramping resumed, taking him to Niagara Falls, where he was arrested for vagrancy and jailed for thirty days, then to Washington, New York, and Boston. In September, he began his return trip—hopping trains and tramping his way west all the way from Montreal to Vancouver—then by ship back to Oakland.[4]

His time as a tramp had a profound effect upon London; he credited these travels with influencing both his literary style and philosophical outlook. He also wrote extensively about his tramp experiences in various forms, including short stories, essays, and memoirs. In 1907, he published *The Road*, which is both a travel narrative and a memoir of his days as a tramp. In it, London attempts to present an insider's perspective of the tramp's world and a more sympathetic understanding of the tramp than that which dominated public opinion at the time. He also cleverly uses the conventions of earlier travel narratives and travel fictions to develop a powerful socioeconomic critique of American society. In doing so, he often portrays the tramp as a *pícaro*, or wandering rogue, contributing to an ongoing process that would, in time, transform the public image of the tramp from a social deviant into a sympathetic and iconic American folk hero.[5]

PRECURSORS TO *THE ROAD*

By the time London published *The Road* in 1907, there was already a substantial body of literature about the American hobo—much of it sociological in nature, focusing on the tramp as a social problem. In general, tramps and hobos were viewed in extremely negative terms:

> Hobos broke with official conventions of social, economic, and religious life and, like all homeless, were seen as pathologically dangerous, spiritually bankrupt, and harmful to their communities. Their poverty was explained as self-imposed rather than caused by economic or political conditions. Hobos often were viewed as suffering from 'wanderlust' or other moral depravities and kept at arm's length, they were ostracized as voluntary dropouts from the working class rather than recognized as a vital subculture within it. (Lennon 17)

Representations of the tramp in newspapers and magazines reinforced this negative view: "In popular representations of these figures, then, they were to be laughed at as lazy buffoons . . . or [viewed] as predatory social deviants with animalistic qualities who must be 'put down' to protect the nation's women and children" (Lennon 18). Cresswell elaborates on this point: "As well as appearing in editorials and learned papers, the tramp was a favorite figure for cartoonists and illustrators. He appeared either as a foolish clown or a hostile threat and sometimes as both at once" (95).

This widespread public perception was, of course, far from the truth. Many tramps had no desire to be on the road but were forced to take to it out of economic circumstances. Cresswell observes that although the tramp was consistently represented as "a shiftless, idle hooligan who avoided work at all costs," in reality, "The vast majority of tramps, however, did work. Many were employed in a number of different industries in the course of a single year, including construction, fruit picking and mining. Tramping changed seasonally as people followed the availability of work" (40). London, who kept a diary as he tramped across the country, confirms this in the following entry: "Honesty of a great majority of hobos, & good heartedness. Good Counsel. The majority willing & eager to work. Gathering in groups & discussing the chances in different parts of the country" (Etulain 58). Moreover, many of these itinerant workers who traveled from job to job saw the hobo "lifestyle" as a way to preserve some degree of autonomy and control over their labor. Lennon contends that many hobos, "embraced their mobility and viewed it as a freedom from the exploitative labor found in the factories. . . . the ability to move allowed some men to view work in alternative and resistive ways. Jobs could be seen as continually temporary due to an intense desire on the worker's part to escape the perils of the industrial labor market" (66). Too poor to afford the price of a ticket that would allow them to travel within the passenger cars of the trains, many tramps resorted to the incredibly dangerous practice of jumping onto moving trains and then literally inserting themselves into the locomotive's interstices—in boxcars, on the narrow platforms between cars, on the car roofs, or on the metal rods beneath the cars. A fall from any of these positions could mean serious injury or death. Yet, in this risky method of transportation, we see how the tramp resists and outwits the economic system set up to exclude him. As Mark Simpson remarks, "By riding the rods, the tramp becomes a parasitic passenger, exploiting the networked system from within, extracting use value from the mass transit he cannot otherwise afford" (112). Marginalized and demonized by the dominant culture, the tramp existed in a world fraught with great risk and had to devise creative methods to survive.

A brief review of the terms vagabond, vagrant, tramp, and hobo are both instructive and revealing. As discussed, the terms *vagabond* and *vagrant* originated in Europe and were brought over to colonial America by the first colonists, and prior to the 1870s, they were the terms most used to refer to the traveling poor. With the economic crisis of 1873, the term tramp entered public discourse and was followed by the term hobo, which, according to Lennon, entered the national lexicon about 1890, although the word had probably come into existence at the conclusion of the Civil War and, prior to 1890, was a Western regional colloquialism used to refer to a migrant worker, as shorthand for the "man with a hoe" or "hoe boy" (14–15). Significantly, by the end of the nineteenth century, all four terms had become synonymous and interchangeable in public discourse, almost always with negative connotations. However, not surprisingly, this was not the case among the traveling poor themselves: those who traveled from job to job thought of and referred to themselves as hobos, carefully distinguishing themselves from the non-working tramp.[6] These questions of terminology are not purely academic, of course, for once something is named, it can then be defined and regulated, and the invention of the term tramp to describe the sudden increase in the mobile poor during the 1870s is a powerful reminder of that fact. As Cresswell puts it, "Once tramps have been defined, they can be dealt with, argued over, confined, made to work on log piles and disenfranchised" (49). He notes that New Jersey enacted the first anti-tramp law in 1876, and by 1886, eighteen other states had done so, as well (49–50).

It's important at this juncture to point out that *tramp* and *hobo* came into existence as gendered and racialized terms. They were commonly and consistently used to refer to white men, even though there were poor migrants of all races and both sexes. The extent to which male hobos of different races mixed and interacted is unclear, although there is plenty of anecdotal evidence to suggest that it was not uncommon for men of different racial backgrounds to travel together or congregate in hobo jungles. This is certainly affirmed by London's memoir, as I discuss below. On the other hand, as DePastino notes, "nonwhites were not integrated on the road as equals. In their encounters with African Americans, for example, hoboes frequently struck a tone of racial superiority, referring to black migrants with epithets that reaffirmed their own 'whiteness'" (82).

Information regarding women hobos is even more scarce, although there is no doubt that they, too, traveled the rails, often disguised as men or boys. Cresswell estimates that somewhere between 5 and 15 percent of the hobo population was female at any given time. He suggests that their absence in public discourse has to do with the ways in which the very notion of a female tramp unsettled socially-sanctioned ideas about gendered behavior and identity: "The female tramp simply did not fit in. She produced in anxious . . . observers a crisis of the male/female binary and a crisis of category itself. Female tramps were clearly outside the normal spaces and social roles ascribed to women, nor did they fit into the male world of the tramp/hobo. Female tramps were perceived as a community of double outsiders–a neither/nor group on the margins of a margin" (109).

Thus, by the end of the nineteenth century, the tramp and hobo had been well-established in the public imagination as an idle, shiftless, itinerant white male, most often represented as either a lazy beggar or a menacing social deviant. But, while most newspaper and magazine publications consistently represented the tramp and hobo in pejorative terms, there were some notable exceptions. As Cecelia Tichi points out, one of the earliest and most important studies of the tramp, which Jack London read prior to beginning work on his travel narrative, was Henry George's *Progress and Poverty* (1879), which examined the problem of poverty in an era of technologically advanced wealth and recognized the emergence of the tramp in the nineteenth century as an economic phenomenon. Tichi comments, "Most vivid is George's oft-quoted statement, also central to *The Road,* that 'the tramp comes with the locomotive.' The tramp and the beggar, in George's term, are the jobless, and the crisis of joblessness is central to his treatise" (28). Another book which served as an important model for London while he worked on *The Road* was Flynt's *Tramping with Tramps* (1899), a firsthand ethnographic account that, according to DePastino, quickly became "the authoritative text on tramp subculture for a generation" (*Citizen Hobo* 49), and it was Flynt to whom London dedicated his own tramp memoir. Unlike George, who saw the tramp as forced onto the road by economic necessity, Flynt maintained that the tramp was motivated by "wanderlust" and was, as DePastino puts it, an "irresponsible but talented underachiever" (*Citizen Hobo* 49). Significantly, London's tramp memoir would reflect and ascribe to both competing views of the hobo lifestyle.

In terms of literary efforts to represent the tramp, Frederick Feied identifies Bret Harte's "My Friend, the Tramp" (1877) as "perhaps the first American short story to make use of the tramp as subject" (10), and as I will discuss shortly in this chapter, there are striking parallels between Harte's story and the first chapter of London's memoir. The tramp also became a stock character, typically appearing as a villainous or morally bankrupt figure, in many

Gilded Age success narratives, as Christine Photinos has documented.[7] But by far the most famous literary depiction of tramps in American literature prior to *The Road* is found in *Huckleberry Finn*, in the figures of the Duke and the King. Twain was intimately familiar with Harte's work, and there is evidence to suggest that, in his creation of the Duke and the King, Twain was responding, at least in part, to Harte's depiction of the tramp.[8] Moreover, it's important to bear in mind that both Harte and Twain had written many stories early in their careers about working tramps, the unskilled miners and migrant workers that were so essential to corporate industries in the American West. However, with his story "My Friend, the Tramp," Harte introduced an entirely new incarnation of the tramp—the unemployed, itinerant, homeless man.

"MY FRIEND, THE TRAMP"

Set in a New England vacation coastal town, Harte's story opens with the unnamed narrator, presumably one of the vacationers, encountering the tramp during a Sunday morning walk by the seashore:

> I came upon a little hollow, in which comfortably extended on the mosses and lichens, lay a powerfully-built man. He was very ragged; he was very dirty; there was a strong suggestion about him of his having too much hair, too much nail, too much perspiration; too much of those superfluous excrescences and exudations that society and civilization strive to keep under. But it was noticeable that he had not much of anything else. It was The Tramp. (181)

The narrator then makes an interesting admission: "With that swift severity with which we always visit rebuke upon the person who happens to present any one of our vices offensively before us, in his own person, I was deeply indignant at his laziness" (181).

Asking the narrator for a quarter until he can find work, the tramp claims to have traveled east from Milwaukee, where he has a wife and six children. The narrator, however, realizes he met him the previous night, when he gave the tramp food, clothing, and money—the tramp posing as an out-of-work fisherman. When the narrator points out the deception, the tramp then claims that he is, by trade, a brick-maker and that he sincerely wants to find work. Suspecting, correctly, that this is a lie, the narrator offers the tramp work for a day or two. Reacting with "surprise" and "consternation," the tramp, after a moment's hesitation, accepts the challenge. Upon reaching the narrator's cottage, the tramp is given the task of rebuilding a twenty-foot stone wall. As the story continues, the tramp proves himself a skillful manipulator, charming the narrator's servants and even children (under the pretense of instilling in them a strong work ethic) to do the work for him. After being dismissed, the tramp moves on to the neighbors, each time skillfully concocting fabricated histories and conditions, designed to prey upon the sympathies of his benefactors. When the tramp eventually asks a young neighbor lady for money, the narrator finally loses patience with him and, giving him a dollar, tells him to leave and never come back. Time passes, and in the story's final sequence of events, another of the neighbors, a wealthy Boston lawyer, tells the narrator that he deals with tramps as trespassers, firing upon them with a shotgun loaded with powder only, as he did the previous night. Unbeknownst to the lawyer, however, the tramp had come by earlier in the day and borrowed the shotgun to go duck hunting, returning the gun loaded with both powder and shot. Alarmed by the revelation, the two men follow a trail of blood down to the seashore, where they find the tramp's meager belongings "tied up in a handkerchief and a crooked grape-vine stick" (196), and the tramp is nowhere to be found.

Several points about Harte's characterization are worth noting here. Although ragged, dirty, and lazy, Harte's tramp is nevertheless a comic but sympathetic figure, a charming rogue, skillfully adroit at fabricating tall tales that he knows will play upon his listener's sympathies (a quality London will later claim is an essential survival skill for the tramp). He is instantly recognizable to the reader as the stereotypical hobo, right down to the cloth bundle he carries tied to the end of a stick. Also notable is the Boston lawyer's contention that tramps are lawbreaking vagrants that must be dealt with harshly—reflecting the prevailing mood of the era that the tramp posed a menace to society.[9]

Moreover, Harte's story is more complex than it appears at first glance. As I suggested in my summary of the narrative, the story criticizes the wealthy, privileged New England families, as much as it does the tramp. In the narrator's observation that the tramp mirrors his own "laziness" back to him, Harte calls the very term into question. And, by shrewdly setting the story in a coastal town populated by wealthy vacationers from other parts of the region, Harte underscores the fact that, like the tramp, they have come here to rest, not work. Why, then, are the actions of the tramp regarded as sloth while the actions of the vacationers are regarded as well-earned rest or relaxation? As London makes clear in his tramp memoir, tramping is incredibly hard work, physically demanding, and fraught with risk. This reconceptualization of work continues in the story when the tramp manipulates the narrator's children into doing his assigned task for him to teach them "the habits of industry," in essence mocking the famous New England, Puritan work ethic.

Harte's story also focuses our attention on different types of social mobility. As Paul Durica observes in his insightful interpretation, both the tramp and the wealthy vacationers lead migratory lives (85). Indeed, *everyone* in the story is from some other place, and by the end of the story, all but the narrator and the lawyer have left the town for other destinations. Yet, the seasonal migrations of the wealthy vacationers are sanctioned by society, while the travels of the tramp are not. As the Boston lawyer puts it, echoing the tramp hysteria of the period, "There is but one way of treating that class of imposters; it is simply to recognize the fact that the law calls him a 'vagrant,' and makes his trade a misdemeanor. Any sentiment on the other side renders you *particeps criminis*. I don't know but an action would lie against you for encouraging tramps. . . . When a tramp appears on my property, I warn him off. If he persists, I fire on him–as I would on any criminal trespasser" (Harte 194). As noted, the lawyer takes pains to emphasize that he does not load the shotgun with anything more than powder; thus, the story's ending seems to suggest that when the tramp is apparently shot with the gun that he himself had loaded, he alone is to blame. But, of course, this is not true. The tramp is shot and wounded because the social and legal identities imposed upon him allow the lawyer to shoot at him in the first place, an idea best conveyed by the tramp's disappearance at the end of the story. He is always invisible, in the sense that others never see his true self but rather those different personas that he is forced to adopt to survive. As the tramp puts it, "Sometimes I have to deceive, when I can't find work." Thus, even though he presents the tramp as a comic stereotype, Harte uses him to expose social inequities and double standards and critique the mainstream society that regarded the tramp as a nuisance, deviant, or threat. This was a technique that London would also effectively employ in his tramp memoir. Moreover, whereas his literary predecessors regularly presented the tramp as a debased caricature (Twain's characterization of the Duke and the King in *Huckleberry Finn* deploys the same stereotypical elements found in Harte but heightens and exaggerates them in order to render the two men as extremely unsympathetic: they are out-and-out con men, master manipulators, and comic figures bordering on the grotesque), London humanizes the tramp, representing him not so much as a threat to society but instead as a victim of it. Feied and Richard Etulain both credit London with being the first major American author to deal with the tramp in a meaningful way. While this contention is demonstrably untrue, it is without a doubt that London helped to establish a prototype of the tramp that would become a staple of twentieth-century American literature—reappearing in works by Hemingway, Dos Passos, Steinbeck, and Kerouac—to name just a few. Moreover, as Etulain observes, London's tramp writing is valuable for his recognition and awareness that the tramp was the by-product of powerful social and economic changes taking place in American society at the turn of the twentieth century:

> The largest contribution of London's tramp writings is their revelation of some of the cultural and intellectual changes that were seeping through America in the years surrounding 1900. . . . Not all who became tramps did so as a means of escape. As London makes clear, many men were forced into hobo life because they lost their jobs or were not physically able to hold positions available to them. When London dealt with tramps, he was writing about a portion of a society experiencing a closing frontier and searching for a new one. And he was the first American author to treat the hobo as a by-product of a culture gradually moving from frontier status to an urban industrial society. . . . London was opening a new vista through his writing about tramps. He was attempting to relate the rising number of tramps to the social and economic changes circulating through America at the beginning of the twentieth century. (26–27)

JACK LONDON'S *THE ROAD*

The chapters that comprise *The Road* were first published individually in *Cosmopolitan* in the months prior to them being published together in book form in November 1907. They are narrated in a lively voice that captures the youthful zest that London was brimming over with in 1894 at age eighteen. In this memoir and travel narrative, London portrays himself as young, strong, and eager for new experiences and adventures. He clearly relishes the challenges presented by tramp life, and in many ways, the transcontinental journey to the East Coast and then back again became a coming-of-age experience and a testing ground to prove his manhood. Throughout the nine chapters that comprise *The Road*, we see how he delights in the camaraderie and fellowship of other tramps, the pride he takes in outwitting train crews as he rides the rails, and the pleasure he takes in relating his adventures to anyone who will listen. As London puts it:

> Every once in a while, in newspapers, magazines, and biographical dictionaries, I run upon sketches of my life, wherein, delicately phrased, I learn that it was in order to study sociology that I became a tramp. This is very nice and thoughtful of the biographers, but it is inaccurate. I became a tramp—well, because of the life that was in me, of the wanderlust in my blood that would not let me rest. Sociology was merely incidental; it came afterward, in the same manner that a wet skin follows a ducking. I went on "The Road" because I couldn't keep away from it; because I hadn't the price of the railroad fare in my jeans; because I was so made that I couldn't work all my life on "one same shift"; because—well, just because it was easier to than not to. (*The Road* 120)

But, for every idyllic moment London describes, there is a darker side to tramp life that he also makes quite clear, as Kevin Swafford observes in his discussion of the tension between adventure and realism in *The Road*. London notes that tramp encounters with hostile train crews often resulted in serious injury and even death, as did falls from trains. Run-ins with police often ended in prison, as it did for London in upstate New York. And for every moment of fellowship and camaraderie with other tramps, there was also hostility, threats, theft, and violence. As Feied observes, "the harsh reality of hobo life continually intrudes into the narrative" (39).

A good case in point is the chapter "Hoboes That Pass in the Night," which begins as a rollicking adventure with London chasing a mysterious hobo named Skysail Jack from eastern Canada across the continent to Vancouver. The two admire each other's skill as tramps, and each is determined to beat the other to the coast. They track each other's movements through messages left at water towers, which, London notes, function as "tramp directories," as well as through word of mouth from tramps traveling in the opposite direction. When London arrives in Vancouver, he learns that Skysail Jack has beat him to it and boarded a ship headed for points unknown. London remarks, "Truly, Skysail Jack, you were a tramp-royal, and your mate was the 'wind that tramps the world.' I take off my hat to you. You were 'blowed in the glass' all right" (101).[10] At this point, London takes the opportunity to offer some insights into and explanation of hobo argot and customs, blending a bit of anthropology into the mix of travel narrative and memoir. The remainder of the chapter, however, recounts his travels in the company of another tramp called the Swede and emphasizes the hardships of tramp life in a way that stands in stark contrast to the high-spirited adventures of the chapter's opening pages. London and the Swede continually struggle to find food and are run out of each town they visit, usually by the local police. But, their biggest struggle, by far, is against the elements, in particular the bitter-cold temperatures during a winter passage from California through the deserts and over the Rockies:

> It was early in the year, and it was cold in those upland pastures. Snow lay here and there on the level, all the mountains were shrouded in white, and at night the most miserable wind imaginable blew off from them. It was not a land in which to linger. And remember, gentle reader, the hobo goes through such a land, without shelter, without money, begging his way and sleeping at night without blankets. This last is something that can be realized only by experience. (104)

Passages such as these make clear that London viewed his narrative as an opportunity to educate his readers about a community existing on the margins of mainstream society, a subculture that he felt was largely misunderstood. As their journey continues, London and the Swede eventually join up with the rear guard of Kelly's Army, which, although providing them with a brief respite from the elements, also subjects them to an odd sort of captivity. Since none of the communities along the railroad line want hundreds of hoboes wandering through their town, they supply the tramps with food but send them on their way without letting them disembark, as the police closely watch their every move at each way station. Once the train stalls at Council Bluffs, Iowa, the hobos, some 2,000 strong, are herded together by the police and held in a closely guarded encampment. London and the Swede slip away from the group and break into an abandoned saloon. Their shelter is cold and drafty, and they are soaked to the bone from the rain outside. London describes the night as the most miserable one he had ever spent in his life. In the morning, as the two men say goodbye and part ways, the Swede mutters, "Never again," over and over. "Never again what?" inquires London. "Never again a hobo," replies the Swede. "Never again a hobo. I'm going to get a job. You'd better do the same. Nights like this make for rheumatism." As they part ways, the chapter concludes with London musing, "Here's to you, Mr. Swede, wherever you are. I hope you got that job" (119).

To reiterate, as London's narration of events progresses, the high-spirited adventure of the chapter's opening pages is quickly overtaken by the almost unbearable hardships of tramp life, in a way that drains tramping of any romantic notions. The recounting of London's encounter with Kelly's Army makes clear that the protesters are under careful surveillance by the police and leaves little expectation that their protest will be successful. Finally, defeated and beaten by the hardships of tramp life, the Swede's assertion that he is going to leave tramping behind and get a job, with 2,000 unemployed men camped just miles away, during the worst economic depression the country had yet seen, strikes both London and the reader as nothing more than wishful thinking. Thus, while superficially, London's tramp memoir may appear to be a coming-of-age adventure story in the life of the country's most successful writer of adventure stories, in actuality, it is also a quasi-ethnographic and sociological account that reveals a tramp subculture existing on the margins of society, inextricably wedded to the economic vicissitudes of an out-of-control capitalist system.

Nowhere does London capture these opposing facets of tramp life better than in the third chapter, "Pictures," which begins with London being invited to sit down to a hearty breakfast by two "maiden ladies." While he devours his food, he regales them with tales of his adventures on the road, as well as the adventures of other tramps, which he freely admits to appropriating and presenting as his own (an important reminder that London's travel narrative also contains travel fictions). Significantly, London describes the encounter as a transaction: "It was a fair exchange. For their many cups of coffee, and eggs, and bites of toast, I gave full value. Right royally I gave them entertainment. My coming to sit at their table was their adventure, and adventure is beyond price anyway" (57). Significant because one of the lessons London learned on the road was that there were other ways to make a living than by selling one's physical strength. Moreover, London is speaking here as much about his readers as he is about his breakfast companions, highlighting his awareness of what the reading public expects from a book by Jack London—entertainment and adventure. Much of *The Road* is written to meet those expectations, and his admission of borrowing the tales of other tramps, suggests that he is not above embellishing his own exploits with those of others. In a meta-critical sense, his claim that "adventure is beyond price anyway" is highly ironic, given that it underscores London's keen awareness that his adventure stories were commodities that competed in a literary marketplace (when *The Road* was published, London was the highest-paid author in the United States), as well as being contradicted by the tramp stories themselves, which highlight the steep physical and social price that tramping entails. Clearly, adventure does have a price, both literally and figuratively.

As if to illustrate the latter point, shortly after this enjoyable breakfast, London goes for a swim in the Susquehanna River and returns to find that he's been robbed by other tramps, leaving him only his clothes. Back on his way, he then encounters a group of "American gypsies" who welcome him because, as London puts it, the American hobo is kin and brother to the American gypsy. However, despite the warm welcome, things suddenly take a dark turn. The gypsy chief, angered by something two boys have done, orders them to come forward and proceeds to whip them. When the boys' mother attempts to intervene, she, too, is whipped, mercilessly and violently. London's response is noteworthy, especially within the larger context of the worldview consistently reflected in his literary works. Assessing the situation, realizing he is outnumbered five to one, he concludes that no good can come of attempting to intervene: "Nor did I move, and without shame I say it; though my reason was compelled to struggle hard against my natural impulse to rise up and interfere. I knew life. Of what use to the woman, or to me, would be my being beaten to death by five men there on the bank of the Susquehanna?" (62). Nevertheless, after witnessing more of the woman's torment, London can hold back no longer. As he rises to go to her aid, one of the other men in the group restrains him and indicates that he is not to intervene. Outnumbered, London can only watch helplessly while the woman is whipped. Finally, he concludes that such is the way of the world—unjust and violent—and that nothing can be done about it.

These rationalizations on his part, whether cowardly or merely pragmatic, are significant given that, in his fiction, London often portrayed the world in Darwinian terms, where the strong preyed upon and took advantage of the weak and where only the fittest survived. *The Road* offers the reader much insight into how formative these tramp experiences were in shaping that worldview, when he was still a teenager. For example, his first tramp experience at age sixteen was spent in the company of other boys close to his own age, whom he calls road kids. They traveled in a closely knit group and, as London describes it, frequently ganged up on lone individuals and beat and robbed them. London likens them to a pack of wolves and just as savage: "Road-kids are nice little chaps—when you get them alone and they are telling you 'how it happened'; but take my word for it, watch out for them when they run in pack. Then they are wolves, and like wolves are capable of dragging down the strongest man. At such times they are not cowardly. They will fling themselves upon a man and hold on with every ounce of strength in their wiry bodies, till he is thrown and helpless. More than once have I seen them do it, and I know whereof I speak. Their motive is usually robbery" (129).

In addition to deepening his view of the world as a violent, predatory place, London's tramp experiences allowed him to witness and experience firsthand the abuses of a merciless, industrial, capitalist system and a corrupt law enforcement and legal system. So that, in the case of the former, as London noted in numerous essays, men, women, and even children were put to work making money for others and, when their strength was robbed of them, were chewed up and spit out, often ending up on the road as tramps. His realization that only a small percentage of tramps were tramps by choice was a turning point in London's intellectual and political development, setting him on a path that would eventually lead him to become a passionate advocate of labor unions and socialism.

London often claimed that he had no motive for taking to the road other than a carefree wanderlust; yet, in fact, his decision to tramp across the country appears to have been motivated by an attempt to break free of this exploitative, capitalist system. When his stepfather became disabled, fourteen-year-old Jack London became the family breadwinner (DePastino xiii). In the autobiographical essay "What Life Means to Me" (1906), London observes, "From then on I was mercilessly exploited by other capitalists. I had the muscles, and they made money out of it while I made but a very indifferent living out of it. I was a sailor before the mast, a longshoreman, a roustabout; I worked in canneries, and factories, and laundries; I mowed lawns and cleaned carpets, and washed windows. And I never got the full product of my toil" (Etulain 139). By age eighteen, feeling increasingly dissatisfied and trapped in a cycle of unrewarding labor, London decided to take to the road and become a tramp.

A good portion of the initial stages of his journey east was made in the company of members of Kelly's Industrial Army of the Unemployed, which, as noted in the preceding pages, was traveling to the nation's capital to protest the lack of jobs available during the economic downturn. However, despite taking to the road at least in part because he felt he was being exploited by a corrupt economic system, while London traveled with these protesters, he was not committed to their cause, even going so far as to steal supplies intended for the larger group, before eventually going his own way. Significantly, London justified the theft by way of the capitalist philosophy he would soon become a vocal critic of: "While we were ahead, skimming the cream, and while the commissary was lost far behind, the main Army, coming along in the middle, starved. This was hard on the Army, I'll allow; but then, the ten of us [his fellow embezzlers] were individualists. We had initiative and enterprise. We ardently believed that the grub was to the man who got there first, the pale Vienna [coffee] to the strong" (139).

Clearly, at this point in his life, London was still immature, self-centered, and naive—supremely confident in his health and strength as guarantors of his "individualism," as he called it. However, as he continued his journey toward the East Coast, his views began to change, as he reveals in the 1903 essay "How I became a Socialist":

> On rods and blind baggages I fought my way from the open West, where men bucked big and the job hunted the man, to the congested labor centers of the East, where men were small potatoes and hunted the job for all they were worth. And on this new blond-beast adventure I found myself looking upon life from a new and totally different angle. . . . I found there all sorts of men, many of whom had once been as good as myself and just as blond-beastly; sailor men, soldier men, labor men all wrenched and distorted and twisted out of shape by toil and hardship and accident, and cast adrift by their masters like so many old horses. I battered on the drag and slammed back gates with them, or shivered with them in box cars and city parks, listening the while to life histories which began under auspices as fair as mine, with digestions and bodies equal to and better than mine, and which ended there before my eyes in the shambles at the bottom of the Social Pit. (Etulain 99)

Realizing that he was but an injury, illness, or employer's caprice away from joining them, London determined to search for an alternative to the industrial, capitalist system, even though it would be years before he fully understood the change in his worldview that occurred as a result of his tramp experiences. As he puts it:

> I think it is apparent that my rampant individualism was pretty effectively hammered out of me, and something else as effectively hammered in. But, just as I had been an individualist without knowing it, I was now a Socialist without knowing it, withal, an unscientific one. I had been reborn, but not renamed, and I was running around to find out what manner of thing I was. I ran back to California and opened the books. I do not remember which ones I opened first.

> It is an unimportant detail anyway. I was already It, whatever It was, and by aid of books I discovered that It was a Socialist. (Etulain 100)

Thus, London may have initially been drawn to the road by a desire for adventure and wanderlust, but the experience did provide him with an introduction to sociology, a new way of thinking about the world. In other words, he did not take to the road to "study" the tramp, but the experience inevitably caused him to reflect upon the social forces that governed the tramp's world. It's also worth noting, as DePastino and others have pointed out, that London kept a diary while he tramped across the continent and back again, suggesting that, on some level, the future writer sensed that it was important to keep a record of his experiences: "Already, it seems, he was attempting to transform experience into story and to become an observer of the subterranean life in which he was also by necessity a participant" (DePastino, *The Road* x).

As Tichi observes, one of the books that had a profound impact on London upon his return to Oakland was George's *Progress and Poverty*, an influential treatise on capitalism which she credits with inspiring some of the "most substantial challenges to industrial capitalism during the Gilded Age" and which more than likely influenced London's own critique of industrial capitalism in his travel narrative (29). Tichi characterizes George's text as "a dense, exhaustive analysis that produced a socio-economic solution focusing on a proposed tax to be imposed on those who gained from others' improvement of land. He urged that the tax revenue be recirculated back to . . . the workers who built America" (28). As London read more widely, he became increasingly convinced that the tramp was largely the result of an exploitative economic system. In his essay "The Tramp" (1904), London articulates this thesis succinctly:

> There being more men than there is work for men, a surplus labor army inevitably results. The surplus labor army is an economic necessity; without it the present construction of society would fall to pieces. Into the surplus labor army are herded the mediocre, the inefficient, the unfit, and those incapable of satisfying the industrial needs of the system. The struggle for work between the members of the surplus labor army is sordid and savage, and at the bottom of the social pit the struggle is vicious and beastly. This struggle tends to discouragement, and the victims of this discouragement are the criminal and the tramp. The tramp is not an economic necessity such as the surplus labor army, but his is the by-product of an economic necessity. The road is one of the safety valves through which the waste of the social organism is given off. And *being given off* constitutes the negative function of the tramp. (Etulain 135)

Thus, London's tramp experiences in 1892 and 1894 were the catalyst for a sophisticated socioeconomic critique that he developed over the intervening twelve-year period prior to the composition of *The Road*, a critique that he carefully incorporated into his tramp memoir. The opening chapter of London's travel narrative, which has striking parallels to Harte's "My Friend, the Tramp," provides an excellent case in point. In it, he describes arriving in Reno, Nevada, in the summer of 1892 and finding it overrun with hobos, all of whom were searching in vain for handouts of food and money. London, too, is turned away empty-handed from dozens of houses and describes having doors slammed in his face, being threatened with calls to the police, or simply ignored. He then observes:

> It began to look as if I should be compelled to go to the very poor for my food. The very poor constitute the last sure recourse of the hungry tramp. The very poor can always be depended upon. They never turn away the hungry. Time and again, all over the United States, have I been refused food by the big house on the hill; and always have I received food from the little shack down by the creek or marsh, with its broken windows stuffed with rags and its tired-faced mother broken with labor. Oh, you charity-mongers! Go to the poor and learn, for the poor alone are the charitable. They neither give nor withhold from their excess. They have no excess. They give, and they withhold never, from what they need for themselves, and very often from what they cruelly need for themselves. A bone to the dog is not charity. Charity is the bone shared with the dog when you are just as hungry as the dog. (16–17)

To drive his point home, in one of the narrative's most effective and entertaining anecdotes, London goes on to describe his attempt to get something to eat at the home of a prosperous man, whose "prosperity had . . . bred resentment against his less fortunate brothers" (18). As the man eats his evening supper, he accuses London of not being willing to work for his food and challenges London to come to work for him in the morning, removing bricks at a burned-out building. London admits to the reader that what he wanted was a free meal before he hopped the evening train back to California, but he bluffs and tells the wealthy man that he's willing to work for him and repeats his request for food: "'You see, I am now hungry. . . . Tomorrow morning I shall be hungrier. Think how hungry I shall be when I have tossed bricks all day without anything to eat. Now if you will give me something to eat, I'll be in great shape for those bricks'" (19).

His prospective employer retorts, "'If I gave you something to eat now, I'd never see you again. Oh, I know your kind. Look at me. I owe no man. I have never descended so low as to ask any one for food. I have always earned my food. The trouble with you is that you are idle and dissolute. I can see it in your face. I have worked and been honest. I have made myself what I am. And you can do the same, if you work and are honest'" (20).

"'But if we all became like you,'" London replies in a withering response, "'allow me to point out that there'd be nobody to toss bricks for you'" (20).

It is moments such as these, Tichi observes, that make *The Road* such a subversive text, since London's intended audience are members of the same prosperous middle class as the "honest man" whom he skewers in the episode—readers weaned on the same bootstrap mythology that London's travel narrative challenges at every turn (Tichi 29–30). Like his predecessor Bret Harte, London uses the presumed sloth of the tramp as a device to enable a critique of the wealthier social classes and *laissez-faire* capitalism. In another example, London recalls sneaking aboard a Pullman passenger train, into a private car, and asking an "itinerant millionaire" for a quarter, juxtaposing the privileged mobility of the wealthy with the dangerous and difficult travels of the tramp.

Nevertheless, these episodes also appear to reinforce the popular perception of the tramp as a lazy bum, unwilling to work, and as a deceptive confidence man—although London makes clear this is a necessary survival tactic: "For know that upon his ability to tell a good story depends the success of the beggar. First of all, and on the instant, the beggar must 'size up' his victim. After that, he must tell a story that will appeal to the peculiar personality and temperament of that particular victim. . . . The successful hobo must be an artist" (21–22).

If we replace the word "artist" with "performer," we can better see such deception as a strategic response to a socially-imposed identity. When tramp and hobo become synonymous with vagrant, bum, con man, and social deviant—imposed identities which are then criminalized—what choice does the tramp have other than to enact or adopt a more socially-acceptable one, to create a more socially-acceptable narrative? To illustrate, London recounts how he acquired food and clothing from a kindhearted woman by lying, leading her to believe that he was a poor orphan traveling to go live with his sister in Salt Lake City. London reflects, "I hope that woman in Reno will read these lines and forgive me my gracelessness and unveracity [sic]. I do not apologize, for I am unashamed. It was youth, delight in life, zest for experience, that brought me to her door. It did me good. It taught me the intrinsic kindliness of human nature. I hope it did her good. Anyway, she may get a good laugh out of it now that she learns the real inwardness of the situation" (29).

Surely, London is being disingenuous here. In *The Road*, human nature is just as likely to be intrinsically cruel, as kind, and it is in that sense that the narrative reveals "the real inwardness of the situation" of life on the road for these marginalized members of society. London replaces the stereotype of the tramp as social menace and deviant with the older stereotype of the *picaro*, the wandering rogue, and the conventions of the picaresque narrative become an integral part of his socioeconomic critique. Each of the nine chapters that comprise *The Road* contains some sort of commentary about the capitalist system that London passionately believed was unjust and exploitative. In "Holding Her Down," he points out that railroad crews had no qualms about hurling tramps to their deaths by throwing them from moving trains, actions backed by the powerful railroad corporations who employed them, as well as law enforcement. After describing the whipping of the gypsy children and their mother in "Pictures," London pointedly compares the abuse to that of child labor in the nation's factories. And, during his thirty days in prison for vagrancy, London notes that prison society replicated the worst abuses of the exploitative, capitalist one outside its walls. As DePastino puts it:

> *The Road* uses the picaresque—a narrative centered on the seemingly random wanderings of a roguish hero—to achieve a critical distance from the mainstream culture. For London, the writer, the road offers a vantage point from which to survey the pretensions, the hypocritical morality, the stultifying manners and the reflexive violence of the propertied classes. Viewing the dominant culture from the bottom up and the outside in, London satirizes such ridiculous refinements as eggcups and Melba toast and exposes such official corruptions as judicial graft and police brutality. From the perspective of the boxcar, the courthouses and parlors of the respectable appear deviant and grotesque, while the hobo emerges as a self-reliant and full-blooded refugee from a despoiled social order. (xi)

As DePastino suggests, London's days as a tramp not only shaped and influenced his belief that industrial capitalism was an abusive and unjust economic system but also led him to develop a similar critique of the legal system in the United States, in particular, law enforcement and the courts. While traveling through New York, London wanted to see Niagara Falls and so hopped a freight train and traveled there. Upon arrival, he walked immediately from the train yard to the falls, joining the other tourists, and transfixed by the sight, he gazed upon it until nightfall. As it was a warm summer night, London walked into the countryside and fell asleep in a field. The next morning, as he strolled back to the falls for another look, he was stopped by a policeman and arrested for vagrancy. Taken before a magistrate, along with other tramps, London was sentenced to thirty days in the Erie County Penitentiary. The court proceedings, as he depicts them, were farcical: No legal representation was given to the accused, and London takes pains to point out that many of the men arrested and jailed along with him for vagrancy had lost their jobs and were searching for work; in other words, they were not tramps by choice but instead tramps by necessity. London describes the conditions in the penitentiary as hellish. The men are fed only bread and water, forced into hard labor, and beaten. Prisoners are encouraged to curry favor with the guards to improve their own situation, and those who do gain the guards' favor exploit their privileged status by abusing and beating their fellow inmates—London himself admits to participating in these beatings as a way to maintain his privileged status as a "hall man." Initially outraged by what he believes is false imprisonment and a miscarriage of justice, London's indignation quickly vanishes as his spirit crumbles under the brutal conditions. He concludes that the legal system and law enforcement in the United States is not only corrupt and unjust but also so powerful as to be unstoppable:

> I saw with my own eyes, there in that prison, things unbelievable and monstrous. . . . My indignation ebbed away, and into my being rushed the tides of fear. I saw at last, clear-eyed, what I was up against. I grew meek and lowly. Each day resolved more emphatically to make no rumpus when I got out. All I asked, when I got out, was a chance to fade away from the landscape. And that was just what I did do when I was released. I kept my tongue between my teeth, walked softly, and sneaked for Pennsylvania, a wiser and a humbler man. (82)

Significantly, London viewed the horrors of prison society as products of the socioeconomic philosophy that governed the world outside its walls: "The whole prison was covered by a network of lines of communication. And we who were in control of the system of communication, naturally, since we were modelled [sic] after capitalistic society, exacted heavy tolls from our customers" (92). Clearly, London's experiences as a tramp were the catalyst for the development of a socialist worldview and an advocacy of social justice, along with a critique of capitalism and government corruption—attitudes that undergird some of his more important later works, like his dystopian novel *The Iron Heel* or *Martin Eden*.

London's tramp days also were influential in his development as a writer. In his relatively short career, which lasted about twenty years (London died when he was forty), London turned out an astonishing amount of work at an incredibly fast pace. This rapid literary output, to a large degree, was driven by London's mounting debts and, as numerous critics and biographers have remarked, account for the uneven quality of his work. However, in his tramp memoir, London attributes his later success as a writer to the necessity of being able to compose vivid and convincing travel fictions quickly, in order to survive on the road. And, he adds, it made him a realist:

> I have often thought that to this training of my tramp days is due much of my success as a story-writer. In order to get the food whereby I lived, I was compelled to tell tales that rang true. At the back door, out of inexorable necessity, is developed the convincingness and sincerity laid down by all authorities on the art of the short-story. Also, I quite believe it was my tramp-apprenticeship that

> made a realist out of me. Realism constitutes the only goods one can exchange at the kitchen door for grub. (22)

This reflection on London's part should serve as a reminder that despite his socialist views, he was keenly aware that stories were commodities that competed in a literary marketplace and that to be successful they had to appeal to a mass audience. Thus, we need to ask ourselves, what do London's narrative choices reveal? What does he emphasize, and what does he omit?

The Road is a problematic text, intriguing as much for the aspects of tramp life it avoids discussing as for those it emphasizes. In his excellent introduction to the 2006 reissue of *The Road*, DePastino notes that while homosexuality was common among the tramp community, the topic is almost entirely absent from London's discussion of tramp life. And although London does acknowledge that there are tramps from all ethnic and racial backgrounds and that he often interacts with them on good terms, his portrayals of African Americans and Chinese Americans are racist caricatures. In addition, an examination of his representations of the tramp reveals a paradox. On the one hand, there is considerable evidence to suggest that London sought to broaden the public's understanding of the American hobo and to depict the tramp in more realistic terms. As his diary entries make clear, he was keenly aware that most tramps were looking for work and traveled from job to job. In an unpublished essay written in 1897 called "The Road," London notes that while there was a segment of the tramp population that resembled the caricature of the dirty, threadbare hobo with a "tomato can on a string"—a stereotype popularized by periodicals of the day and newspaper cartoons—the hobo community was much more diverse. Rather, he continues, the largest segment of that community by far was the "working tramp" or "bindle-stiff"—the migrant worker: "These are the men, who, in New York, travel into the Genessee country to the hops; in the Dakotas, to the harvests; in Michigan, to the berry-picking; and in California, to the vintage, hops, and harvest" (Etulain 72). And, just as the hobo community is diverse, London adds, so, too, are the circumstances that lead men onto the road. Curiously, the working tramp makes only sporadic appearances in *The Road* but, nevertheless, provides some of the most powerful moments of London's socioeconomic critique, as in the case of the out-of-work teamster sentenced to thirty days in prison for vagrancy, when, in fact, he is traveling in search of a job, or, as in the case of the bindle stiffs who are frequently targeted and robbed by road kids and other tramps.

Yet, even as London offered a broader perspective on the tramp—one grounded in a socioeconomic context—than found in other publications of his era, he also frequently represented the tramp in terms that are both romanticized and stereotypical. In *The Road*, he often presented his tramp experiences as youthful escapades and thrilling adventures, and he often represented other tramps as colorful, exotic figures. In part, this had to do with London's decision to adopt the conventions of the picaresque narrative to facilitate his social critique. But to better understand the tensions and contradictions in *The Road*, it is helpful to situate the composition of the text (1906–1907) against the actual journeys that took place in 1892 and 1894. Interpreting *The Road* requires us to consider the passage of time. When he composed his memoirs, London was twelve years removed from the events he was recalling and re-imagining, and this intervening period could account for some of the nostalgic and romantic moments of the text. However, as noted, he kept a diary of his journeys as he was traveling, and this explains the freshness and immediacy of the episodes he recounts.[11] As his breakfast conversation with the two maiden ladies makes clear, he was keenly aware that he was known first and foremost as a writer of entertaining adventure stories. In her excellent essay "Tracking Changes in Jack London's Representation of the Railroad Tramp," which examines *The Road* in relationship to London's earlier tramp writing, Christine Photinos argues persuasively that London presents his tramp experiences as rugged adventures at least in part to satisfy the expectations of *Cosmopolitan*'s editors and readers (180). In addition, we must also keep in mind that London's text falls firmly within that most durable of American literary genres—the American road narrative. Thus, in assessing the work, we need to ask ourselves how generic expectations and conventions might have affected London as he composed his text, as well as the expectations of his contemporary audience. This is precisely the focus of Tichi's insightful essay "Canonizing Economic Crisis: Jack London's *The Road*." In it, she reminds us that when London was composing the memoir in 1906 and 1907, he was one of the most popular and widely read authors of his day, well known for his adventures at sea and in the Yukon: "To the reading public, then, London was a successful, peripatetic brand-name author whose *The Road* was an autobiographical back story from the impecunious but plucky days in his late teens" (22). Thus, Tichi argues, London's readers would be predisposed to greet *The Road* as another of London's adventures, and London would have been keenly aware of that expectation as he worked on the text, thus the abundance of episodes that highlight his physical prowess and narrow escapes. Moreover, in thinking about how to cast and frame his experiences as a youthful tramp, London had ample literary precedent to draw upon, and this explains, I think, why he frames his experiences in accordance with the American road narrative's archetypal themes, in particular, the quest for "personal liberty, self-discovery, identity formation, and a new life of arrangements that are distinctly different and severed from those of the past" (Tichi 21). To be clear, I think that those readers who find these themes in London's road memoir are not mistaken—they are all there in vivid relief. It is so vivid (and this, of course, is Tichi's point) that they tend to obscure or overshadow his socioeconomic critique. To fully appreciate and understand *The Road*, we must, as Tichi encourages us, read against the grain of the powerful expectations of the American road genre and look for those places where it does not conform with them. Such moments, she argues, are evidence that London intentionally embedded his subversive critique of American capitalism within the conventions of the very popular genre of the American road narrative precisely because he knew that his audience would be those readers who would be most resistant to it—well educated, well-to-do, *laissez-faire* libertarians. As Tichi puts it, "Readers must be wooed but simultaneously educated into realities from which they insulate and/or harden themselves. The narrative surface of London's road story is the salable literary property, but just beneath that surface lies the socioeconomic and political case study that belies any notion of Whitmanesque 'loafing' as it exposes an America in crisis" (30).[12] Furthermore, as Adler observes, by adopting this narrative strategy, London was aligning himself with the even older genre of the picaresque novel:

> Middle class interest in tramping as a travel genre is already discernible in picaresque novels of the sixteenth century, which exploited the adventures of unemployed vagabonds and traveling apprentices as an opportunity for critical social commentary. Stories featuring young servants who lived by their wits, frequently changing masters and places, explore the different perspectives yielded by frequent shifts in geographical and social vantage points. The theme of social mobility—more particularly, the exploration of society's "lower depths"—remained an important motif in realist travel literature as well as in travelers' lore. (346)

Adler's astute observation is important, I think, because it reminds us of how deeply interconnected and intertwined travel narratives are with travel fictions and of the long tradition of social commentary the two forms share. She further notes that by the mid-nineteenth century, "literary expressions of the art of travel had come to include an ever stronger aesthetic of travel realism—the life and manners of a country's poor constituting the sights which commanded the serious traveler's attention. The habitats of the 'dangerous classes' were conceived as a new wilderness for middle class exploration, and facing its perils earned merit as worthy travel adventure" (Adler 350). Flynt's *Tramping with Tramps*, a first-person exposé of tramp life, is a perfect example of the type of travel writing that Adler describes. We would do well to remember that not only had London read Flynt's book, but he also dedicated *The Road* to Flynt, calling him "The Real Thing, blowed in the glass." London clearly admired Flynt's account, and it no doubt served as a narrative model for his own book. London had also conducted a similar undercover investigation, posing as an out-of-work sailor, in the city of London's slums, published in 1903 as "The People of the Abyss." By the time he began composing *The Road*, economic and social inequality were very much on his mind.

As Adler's comments intimate, part of the appeal of these travel narratives focusing on the poor and other marginalized members of society was that they satisfied a kind of voyeuristic curiosity among middle- and upper-class readers, and in *The Road*, London skillfully uses that curiosity to his advantage and powerfully images it in the following passage from his opening chapter:

> At other houses the doors were slammed in my face, cutting short my politely and humbly couched request for something to eat. At one house they did not open the door. I stood on the porch and knocked, and they looked out at me through the window. They even held one sturdy little boy aloft so that he could see over the shoulders of his elders the tramp who wasn't going to get anything to eat at their house. (16)

Here, London creates a moment that serves as a metaphor for the entirety of his tramp writings. As he appears to allow his readers to gaze through the window into the tramp's world from the comfort of their own homes, so, too, does London look back upon them from his own critical vantage point. As DePastino remarks, "One must imagine *The Road* as titillating middle class readers as it pulls aside the curtain on a fearsome and inscrutable tramp world. In letting his readers in on this subculture, London gives them a powerful new perspective from which to critique their own sheltered lives. . . . Indeed, in London's book, the road itself is but a microcosm of a larger social world whose grasping brutality leads some to mastery and others to the Social Pit" (*The Road*, xli–xlii).

Thus, the novelty of London's text is not that he used a travel narrative to provide a sophisticated critique of social mobility and migration. As Adler notes previously, that emphasis has often been a common feature of the genre. Rather, the novelty lies in the clever way in which he was able to exploit the evolution of a particular American strand of travel writing, both fiction and nonfiction, one that, for over a century, helped to create and reinforce a popular American mythology built around individualism, self-reliance, and a "bootstrapping" ideology. As Tichi observes, by the time London wrote *The Road*, that mythology and ideology were so wedded to the genre of the American road narrative that London was able to use it as a literary Trojan horse, allowing him access to readers who revered the very mythology and ideology his travel narrative challenges and subverts at every turn.

CONCLUSION

In 1994, one hundred years after eighteen-year-old Jack London embarked upon the transcontinental journey that would radically alter his view of the United States and its economic system, Leslie Marmon Silko published a pair of related essays that reflect upon the dangers of traveling in the United States as a person of color. In "Fences Against Freedom," she writes:

> We citizens of the United States grew up believing this freedom of the open road to be our inalienable right. The freedom of the open road meant we could travel freely from state to state without special papers or threat of detainment; this was a "right" citizens of Communist and totalitarian governments did not possess. That wide open highway was what told us we were U.S. citizens. Indeed, some say, this freedom to travel is an integral part of the American identity. (1996a, 114)

She concludes the essay's companion piece, "The Border Patrol State," with the following recollection:

> One evening at sundown, we were stopped in traffic at a railroad crossing in downtown Tucson while a freight train passed us, slowly gaining speed as it headed north to Phoenix. In the twilight I saw the most amazing sight: dozens of human beings, mostly young men, were riding the train; everywhere, on flatcars, inside open boxcars, perched on top of boxcars, hanging off ladders on tank cars and between boxcars. I couldn't count fast enough, but I saw fifty or sixty people headed north. They were dark young men, Indian and mestizo; they were smiling and a few of them waved at us in our cars. I was reminded of the ancient story of Aztlán, told by the Aztecs but known in other Uto-Aztecan communities as well. Aztlán is the beautiful land to the north, the origin place of the Aztec people. I don't remember how or why the people left Aztlán to journey farther south, but the old story says that one day, they will return. (1996b, 123)

Reflecting on these passages from Silko, I'm reminded of how central migration is to the human experience, but reading her alongside of London's tramp writings, I'm also reminded that while the idea of an absolute right to travel the open road may be a deeply cherished part of our American mythology, it has never been a reality. The road has never truly been open; travel in the United States has always been a privilege, never a guarantee. And the young men Silko witnessed traveling north on the train are a vivid reminder that the economic inequities that drove London and other tramps to ride the rails are still very much present in our nation and in our world. Even as I write this, there is talk of building a wall across our southern border, and immigration remains one of the most divisive issues in the United States.

To delve into and study the origin of the tramp and the literature that helped to create "him" is to confront the contradictions, fears, ironies, divisions, and myths that are inextricably woven together in American history and that, at times, border on the fantastic. It is the story of an economic crisis so powerful that it would, in a country that cherished mobility as part of its national identity, transform the tramp, almost overnight, into one of its most feared and despised members. It is to encounter, in Jack London, one of the most improbable rags-to-riches stories in all American history—a teenage boy who, like so many others, took to the road to avoid being crushed by an exploitative economic system and would go on to become the country's first millionaire author and one of that system's most vocal critics. Who, in writing about his tramp experiences, would excoriate that era's *laissez-faire* capitalism and help to reshape the American public's perception of the tramp and hobo. It is to be confronted with an astonishing lack of information about the female tramp, who even today is more a figure of conjecture and speculation than historical fact, a figure who transgressed so many social boundaries that her very existence was considered outside the realm of possibility. If the tramp survived, in part, by making himself invisible to the authorities and secreting himself, literally, into the interstices of the locomotive and American society, the female tramp had to make herself invisible to other hobos, often by way of cross-dressing and disguising herself as a boy or young man, or run the risk of sexual assault or violence. And finally, it is to be confronted with a subculture that stubbornly and creatively resisted and persevered in the midst of the very society that rejected it, even as it reproduced some of that society's worst abuses.

NOTES

1. See, for example, Frederick Feied, Tim Cresswell, Todd DePastino, Cecelia Tichi, and John Lennon.
2. Lennon goes so far as to insist that train hopping is part of what defines a hobo (13). See also Feied, 9.
3. For a discussion of the etymology of the word "tramp," see DePastino, *Citizen Hobo*, 5, and Cresswell, 48–49.
4. For a thorough summary of London's travels as a tramp, see Richard Etulain's introduction to *Jack London on the Road.*
5. For an excellent discussion of how the tramp evolved into an American folk hero, see Paul Durica's "Flux Americana."
6. See, for example, Lennon, 15.
7. See Photinos, "The Figure of the Tramp in Gilded Age Success Narratives," *The Journal of Popular Culture* 40.6 (2007): 994–1018.

8. Harte and Twain had a famously complicated relationship. Harte had supported Twain early in his writing career and helped to edit Twain's travel narrative *Innocents Abroad*. However, they later had a well-documented falling out over money (Twain claimed that Harte had asked for a loan with no intention of ever paying it back), and Twain went on to become Harte's greatest detractor. For a discussion of how Twain's Duke and King may have parodied relationships depicted in Harte's stories, see Axel Nissen's "A Tramp at Home: Huckleberry Finn, Romantic Friendship, and the Homeless Man," *Nineteenth-Century Literature* 60.1 (2005): 57–86.

9. Lennon notes that by 1886, nineteen states had enacted anti-tramp vagrancy laws (24).

10. According to DePastino, the phrase "blowed in the glass" denotes authenticity, a reference to the practice of glassblowers' blowing seals of authenticity within the glass itself (xxiii).

11. Durica proposes that the form of London's narrative, episodic and filled with digressions, represents a kind of tramp aesthetic, a "textual simulation of the experience of tramping itself" (93).

12. There are other ways to account for the paradoxical qualities of London's memoir, of course. Feied, for example, views this paradox as reflective of the opposing philosophical views which had the greatest influence upon London—those of "Darwin and Nietzsche on the one hand; Marx and Engels on the other" (83). Whatever the case may be, I think it is fair to say that London, ironically, perpetuated stereotypical and romantic notions about the American hobo, even as he sought to challenge them.

Chapter Four

Travel Fictions and Literary Travelers

Jacques Poulin's Volkswagen Blues

French Canadian author Jacques Poulin's novel *Volkswagen Blues* (1984) begins with the sentence, "He was [a]wakened by the meowing of a cat," a playful and ironic gesture given that one of his protagonists, Jack Waterman, who also happens to be a French Canadian novelist, believes that the first sentence of any novel "should always be an invitation that no one could resist" (23). His own novels disappoint him in this regard, as would, one would suspect, the first line of the novel he inhabits, written by his creator. This is just one of many instances in *Volkswagen Blues* in which the narrative calls attention to itself as a textual artifact. These moments of self-reflexivity, of which there are many, emphasize that Poulin is keenly aware of writing a novel that belongs within a larger literary tradition—that of the transcontinental American journey.[1]

The cat belongs to a young Métis woman, Pitsémine (also known as La Grande Sauterelle), who hitches a ride from Jack. Together the three of them will travel from Québec to San Francisco in Jack's old Volkswagen bus in search of his long-lost brother Théo—the first half of their journey retracing the route of the early French explorers, and the second following the route of the Oregon Trail and the nineteenth-century pioneers. They finally do find Théo in San Francisco but in a debilitated state, suffering from dementia and "creeping paralysis." Reluctantly deciding that his presence there will cause Théo more harm than good, Jack gives the Volkswagen bus to Pitsémine, and they part ways. She stays on in San Francisco, and he returns to Québec.

Midway through the novel, Pitsémine's cat disappears, and an entire short chapter is devoted to the search for it. When the cat returns of its own accord and refuses to eat, Jack and Pitsémine conclude that he must have found something to eat on his own and christen him Chop Suey, also the title of the chapter—one that appears to be nothing more than a whimsical digression from the larger story. However, the chapter is actually a very important one if the reader recognizes that Poulin's novel is one that operates by association. One of the most distinctive features of *Volkswagen Blues* is its intertextuality. Every chapter has a reference to another text: a novel, a book, a film, a painting, a mural, a song, a poem. It is also a novel filled with references to historical events and personages. While these artistic and historical references can, at times, seem quite arbitrary and random, each allusion offers a unique interpretive framework for the novel's meditations upon the role and purpose of art, the role and purpose of different types of traveling, and, perhaps most importantly, for its reconsideration of American history, particularly in regard to the settlement of the North American continent and of related, popular American mythologies. The many allusions also serve to remind us that America is as much a textual construction as it is an actual place. Beatrice Guenther aptly describes Poulin's project in the following passage:

> The *lieux de memoire* [sites of memory] in Poulin's *Volkswagen Blues* function in two ways. First they act as a trigger to unearth and decipher lost or effaced histories, and second–in a contradictory manner–as signposts leading to the demystification of one's mythologized origins. This demystification targets both the mainstream, nationalist history glorifying the 'originary' French presence on North American soil as well as an uncritical, idealized elaboration of the dispossessed Amerindian peoples. (68)

So, for example, if one begins to investigate the origins of the Chinese dish chop suey and the mythology surrounding it, we learn that the phrase loosely translates as "mixed pieces," "odds and ends," or "to break into many pieces." The dish itself originated in Toishan, a region of southeastern China that was home to many of the first Chinese immigrants to America in the nineteenth century. However, despite its mainland China origin, popular myths gave the dish a distinctly American genesis, either as an invention of Chinese immigrant cooks working on the transcontinental railway in the nineteenth century, or as a concoction created to appeal to non-Chinese American consumers. *Chop Suey* is also the title of a 1929 painting by Edward Hopper.[2]

How might even these initial findings provide us with an interpretive framework for Poulin's novel? For one thing, "mixed pieces" and "odds and ends" are an apt way of describing the form of Poulin's narrative. It is largely a pastiche, both in the musical sense, as it borrows and mixes motifs from other works, and in the literary sense of imitation—as numerous critics have noted, *Volkswagen Blues* is, among other things, an homage to Jack Kerouac's *On the Road*.[3] Chop suey's origins in the Chinese province that was home to many of the first immigrants to the United States is also significant, as that nineteenth-century mass migration is one of the often overlooked histories of the Americas that Poulin wants to call to the reader's attention. The transcontinental railroad origin myth serves the same function and focuses attention on another important historical event in which Chinese immigrants and Chinese Americans played a crucial role. Finally, as incorporated into Hopper's painting, the chop suey allusion nods toward the ways in which American histories, stories, and mythologies are told, re-told, and reinterpreted by historians, writers, and artists—just as the many works of art referred to in Poulin's novel do and just as Poulin himself does.[4]

The most important function of intertextuality in *Volkswagen Blues* is articulated neatly by Pitsémine: "You shouldn't judge books one by one. I mean, you mustn't see them as independent objects. A book is never complete in itself; to understand it you must put it in relation to other books, not just books by the same author, but also books written by other people. What we think is a book most of the time is only part of another, vaster book that a number of authors have collaborated on without knowing it" (124–25). This is an excellent summation of another of this novel's main points: that the history of the Americas is so complex that it cannot be communicated through a single text and can only begin to be understood by putting its varied histories in relationship to one another.[5] Thus, this is a novel about the problems of historiography: how our understanding of the past is mediated by those texts that present it to us—an enterprise that inherently distorts the past even as it seeks to document it. As Adam Weisman puts it, "*Volkswagen Blues* is also laced through with a dense web of intertextuality that puts it in dialogue both with historical events in the exploration of the North American continent and with the development of an historiography and a mythology surrounding those events" (8).

In the opening chapter, the reader is presented with the first of many clues about Théo's whereabouts, a postcard he sent to Jack years ago, discovered to be connected to a series of other texts that establish important themes in Poulin's novel. We learn that Jack kept the postcard in a book by Walker Chapman about the legend of El Dorado called *The Golden Dream*—reminding the reader of the ways in which the desire for gold drove much of the Spanish conquest of the Americas, as well as the importance of oral histories and storytelling for preserving and transmitting native histories and culture. The postcard also contains what will turn out to be a reproduction of a portion of a manuscript by Jacques Cartier, the French explorer who claimed Canada for France in 1534. When Jack and Pitsémine stop at a museum in Gaspé, Québec, so that La Grande Sauterelle can say hello to her mother, who works as a custodian there, her mother recognizes the text immediately. Thus, Pitsémine's mother is a custodian in a much more important sense than the term implies: she is the custodian of the native histories housed in the museum. She remembers and understands them, and it is this role which her daughter also performs throughout the novel. As Susan Rosenstreich observes:

> The cleaning woman at the museum, which in fact overlooks the site where Cartier erected the cross on July 24, 1534, that signified his claim to Nouvelle France in the name of the French king, is Pitsémine's mother, her Montagnais parent. As a Montagnais, this character brings into the novel the Algonquin-speaking peoples believed to have occupied the territory of present-day Québec as early as 2,000 B.C. Simply the presence of such an individual in the text exposes the unnatural claim of political dominion in Cartier's document. Théo's post card contests in this way the monolithic history recorded in and by documents, at the same time as it provides a clue to his message. (123)

While at the museum, Jack and Pitsémine see two maps of North America set side by side. The first depicts the extent of French colonization of North America in the mid-eighteenth century. The second depicts North America before European contact: a listing of Native American cultures, many of which were wiped out by European conquest and colonization. The startling juxtaposition vividly drives home the extent and scale of destruction caused to Native cultures by European conquest and colonization.

Thus, in the opening chapter, Poulin quickly establishes that this novel is concerned with the history of the Americas, North America in particular, and the ways in which those histories are recorded, interpreted, and reinterpreted. He signals his concern for the impact of European colonization upon Native peoples and how that colonization affects them to this day, and in the partnership between the French Canadian man and the Native American woman, he also alludes to earlier legendary partnerships like Pocahontas and John Smith, and Sacagawea and Lewis and Clark, and the ways in which those stories have become greatly mythologized through their varied retellings. The image of the postcard imprinted with a sixteenth-century French colonial text, carried within a book about the legend of El Dorado, symbolizes the importance of intertextuality to this novel—it is in many ways a novel about other texts, including histories, legends, maps, films, and other novels—but also metaphorically suggests the ways in which the European accounts, presented as monolithic and authoritative, are, in fact, deeply intertwined with traces of suppressed Native accounts that contest and challenge the veracity of European versions of history. These suppressed and forgotten histories of the Americas are precisely what our protagonists will repeatedly encounter as they travel across the continent.

ROOTS AND ROUTES

To borrow a phrase from James Clifford, the route that Jack and Pitsémine follow is closely linked to their roots. As they begin to retrace Théo's route, they learn that it, in turn, retraces the "first French explorations in America" (15) by explorers whom Jack openly admires. La Grande Sauterelle, on the other hand, as she puts it, is "on the side of the people who were robbed of their lands and of their way of life" (17) by the European colonizers. We learn she is of mixed race—her mother is a Montagnais Native American, and her father is French Canadian—and we get a glimpse of the inner conflict this creates for her when she tells Jack that she is "not a real Indian" (17). Later, we learn that when her parents married, they were disowned and ostracized by each of their respective cultures, white and Indian, suggesting that both cultures view intermarriage as a form of cultural betrayal. As a product of this union, Pitsémine feels like an outcast, not belonging fully to either culture. Her challenge, as the journey continues, will be to resolve this identity crisis and make peace with herself. Thus, the two main characters clearly represent European and Native American views, respectively.

Moreover, the routes of the early French and French Canadian explorers and missionaries contest the popular American mythology of the settlement of the North American continent as a seamless, uninterrupted movement from East to West. In a similar fashion, Poulin challenges the popular misconception of the Oregon Trail as a nineteenth-century creation of the white American pioneers:

> When he was heading out west, to the little cove that only he knew about, he always took the Oregon Trail: it was the oldest trail in America. It was older than the history of the conquest of the West, older than the *coureurs de bois* and the pioneers, older than all the emigrants with their ox-drawn carts. It was as old as the Indians and probably as old as America. (174)

VAGABONDS AND EXILES

Poulin's novel also provides different opinions on the act of traveling itself. One view is that the traveler is a vagabond, outside of socially-accepted norms: "To the question regarding occupation, Théo had replied traveler and he was unable to explain what that occupation consisted of and how it enabled him to earn his living" (51). But Poulin also presents us with a romanticized view of the vagabond or tramp, epitomized in the novel by an elderly hitchhiker who resembles Ernest Hemingway: "One way or another he'd been traveling all his life. He liked the road. It was his way of living" (173). "'I'm a rambler, even a tramp, you might say, but I'm not a bum'" (178). An interesting moment occurs during this encounter that situates the romantic notion of the tramp within a larger context. The old man asks our travelers if they know the song "No Roots in Rambling" by Jerry Jeff Walker, and he and La Grande Sauterelle begin to hum and sing the melody: "It was a very sweet, nostalgic country ballad, and Jack managed to catch a few phrases such as: 'I'm now alone and I know I need to ramble' and 'It's the call from deep inside' and 'The blues will haunt me till I die'" (174). This passage reminds us that Poulin frames his entire narrative within the musical tradition of the blues, an African American musical form that has its roots in slavery and the condition of exile. Thus, just as Poulin takes pains to re-introduce the Native American and French Canadian colonial experiences into the larger American history he is examining, so, too, does his title demand a constant awareness of the black American experience within an understanding of American history. The reference to Walker's interpretation of the song as a country ballad reminds us of how the blues have been continually mined to invent and reinvigorate other musical forms such as jazz, country western, and rock and roll, often with the effect of distancing the newer forms from the historical basis of the original blues songs—just as the romanticized retellings of American history referenced throughout the novel obscure or distort its realities, and just as later interpretations of blues songs present the wandering and rambling that is one of the genre's archetypal themes as not so much a condition of an oppressed people but instead as a romanticized, happy, nomadic state.[6]

This myth of the romantic life on the road is, to a certain extent, called into question by Pitsémine, who also leads a nomadic life on the margins of society, but this is not by choice, as in Théo's or the old man's case, but, as alluded to earlier, because her family has been exiled by both communities from which she descends. This contrast between the traveler by choice and the traveler by necessity is one that recurs throughout Poulin's novel in many important ways, all with significant implications. Earlier, I wrote about the way in which the chop suey origin myth was rooted in the larger story of Chinese immigrants and contracted workers who provided the cheap labor to do the difficult work of constructing the transcontinental railroad, which, when completed, became the conveyance of the privileged, wealthy traveler. Moreover, Poulin also reminds us of the impact the railroads had upon American Indian plains cultures:

> But instead of falling asleep again, [Pitsémine] launched into a violent outburst against all the whites who had turned buffalo hunter for financial reasons, like Buffalo Bill or the employees of the fur trading companies or for sport like the rich people who came out from New York or Washington on special trains and without even stepping outside, unloaded their rifles into herds that

> beaters had pushed along the railway line; her most unflattering epithets were reserved for the Grand Duke Alexis of Russia, son of Czar Alexander II, who had come to hunt buffalo, complete with his retinue and his cases of champagne in 1872. (127–28)

And although there is never any direct mention of the Long Walk, the forced march the Navajos were made to undertake in 1864, there is a reference to Kit Carson, who was to a great degree responsible for this dark chapter in Navajo history. Carson is just one of many of Jack's personal heroes whose legendary status Pitsémine will challenge over the course of the journey.

READING, WRITING, AND TRAVELING

As the search for Théo continues, Jack and Pitsémine learn that he was arrested in Toronto on a weapons charge and, at the time of the arrest, was traveling with two books—*On the Road* and *The Oregon Trail Revisited*. The references to these two works, one literary and the other historical, are another important key to understanding Poulin's novel. Both of Théo's books are archetypal models of transcontinental American journeys. The reference to Kerouac's novel suggests the recreation of a famous literary model of travel but in a way that self-consciously acknowledges that *Volkswagen Blues* belongs to a particular literary tradition and genre while also acknowledging Poulin's literary predecessors, who helped to establish that genre's conventions. The reference to *The Oregon Trail Revisited*, a history of the nineteenth-century pioneers who traveled it on their journey west, becomes both a guide to the trail and to the history of the trail. Jack and Pitsémine use the book as a kind of road map while they simultaneously imagine the stories of the pioneers contained in it as a *tableaux vivant*, in which Jack imagines Théo as a heroic wagon train guide. Nor is it a coincidence that one of the two books is a literary work and the other a historical one—it is Poulin's way of reminding the reader that "America" is as much a literary and a historical construct as it is a geographical place. And, as Rachel Adams suggests, the reference to Kerouac, who was French Canadian, reminds us of the important Francophone influence in American culture.[7]

The activities of travel and reading are closely linked in Poulin's novel. He reminds us that, like navigation, reading is an act of interpretation. So that, just as Pitsémine, who is normally an unerring navigator when she is on the road, seems clueless when it comes to understanding the urban terrain of inner-city Detroit, the novel is also filled with moments of startling interpretations of the texts, works of art, and monuments that our protagonists encounter during the course of their journey, reminding us that, like travelers, readers have extremely varied reactions to the artistic terrain they are moving through or encounter. For example, while in Toronto, Jack and Pitsémine gaze upon the architectural achievement of the Royal Bank, which, according to brochures they have read, has windows dusted in gold—all of which reminds Jack of the legend of El Dorado: "The building looked as bright and warm as honey, and they couldn't help thinking about the gold of the Incas and about the legend of Eldorado. It was as if every dream was still possible. And for Jack, in his heart of hearts, it was as if all the heroes of his past were still heroes" (55). While Jack finds this comforting, the astute reader notes how the legend of a powerful Native American ruler is here associated with the economic power of a major financial institution: capitalism has replaced the Inca king as the seat of absolute power. It is hard, in a "too big to fail" era, to find Jack's analogy as optimistic and comforting as he does. And while the gold dust of the bank's windows reminds Jack of the beauty of the image of the Inca king dusted in gold and glittering in the water of the lake, it reminds the reader of the lust for gold that drove both the Spanish conquest of the Americas and the gold rush across the American plains, each producing myriad conflicts, massacres, and atrocities. As Roger Hyman puts it, "Here the religious purity of the original El Dorado . . . has become finally the corporate capitalism of the gold-dusted Royal Bank Plaza, a comment about the way in which the wealth of this continent is quite literally built on the wealth of those whose land this once was" (12).

A similar moment occurs during a visit that Jack and La Grande Sauterelle make to the Detroit Institute of the Arts. Here, they view Diego Rivera's mural depicting the auto manufacturing industry in Detroit:

> The room was nearly ten metres high, and it was lit by the natural light that came in through the glass roof. Rivera's work covered all four walls. In tones dominated by pale green, pale yellow and, above all, grey, it depicted huge industrial machinery with blank-faced workers bustling around it. . . . All the faces were immutably serious, solemn almost, and their gravity added to the impression of austerity emanating from the drab colours. The overall effect was one of heaviness, sadness and exhaustion. (65)

As they prepare to leave, they notice a small red car in the mural and later conclude that "the red car was a symbol for happiness and that Rivera had wanted to express something very simple: happiness is rare and getting it requires lot of trouble, fatigue, and effort" (68–69). Pitsémine adds that the same idea is found in a song by George Brassens, taken from a poem by Aragon: "There's nothing a man can count on for sure / Not his strength nor his weakness nor his heart / And when he believes that he's opening his arms / His shadow is that of a cross" (69).

This is an interesting moment in the narrative, particularly from the standpoint of interpretation. The somber austerity of Rivera's mural clearly has a powerful effect upon both Jack and Pitsémine, yet the apparently straightforward and "simple" meaning of the murals they arrive at is extremely reductive and appears to overlook much more complex ideas about the relationship between labor, capitalism, and industrialization. As Hyman remarks, Jack and Pitsémine ignore "the huge sweep of the mural, and its deep suspicion of capitalism. . . . Their simplistic conclusions about the Rivera mural have nothing whatever to do with Rivera's work or his social analysis (where the 'red car' can be seen as a Marxist symbol of the exploitation of labour)" (10). And yet, however reductive their interpretation of the murals may seem to the reader, their attempt to understand the murals by considering them in relationship to other texts seems entirely in line with Poulin's ideas about all texts being inseparable from others. So the episode provides, first of all, another example of the unpredictability of interpretation (I hesitate to say misreading) and, secondly, another example of the sort of artistic re-interpretation that I would argue Poulin sees as both obfuscating history and, paradoxically, revitalizing it, as artists borrow from one artistic form to give life to another. Finally, the Rivera mural is the most visible example of recurring socialist politics that is woven through Poulin's narrative. In addition to references to Jack London, who was influenced by Marx and was an early proponent of labor unions, the novel's epigraph is taken from *Ringolevio*, a novel by Emmett Grogan, who, as Hyman notes, was a founder of the San Francisco Diggers, an important anarchist group of the 1960s:

> . . . whose anarchic politics of the free stimulated much of the hippie movement in North America. Their name was taken from the equally short-lived agrarian communists of the late 1640s who opposed Cromwell's Enclosure Laws and the assumptions of private property behind them by cultivating what had been common land in Surrey. Their anarchic social politics was, as was the case with their modern successors, aimed at the fundamental assumptions of western capitalism, in particular at the concept of ownership, the same assumptions, one notes, that underpin the destruction of much of the world's indigenous cultures, and which are echoed in the structure and characters of *Volkswagen Blues*. (4)

Poulin also foregrounds his interest in problems of textual interpretation through his two main characters, who represent two very different types of readers:

> While La Grande Sauterelle devoured every book she could get her hands on, Jack Waterman was an anxious, parsimonious reader. He had his favorite authors, all of whose books he had read, but those authors were few in number . . . And he had his favorite books, which he reread frequently, and which

> were like old friends. In general, he felt no need to read during the periods when he was writing a novel; at those times, which were lasting longer and longer, other people's books made him impatient and sometimes even jealous. (27)

These two types of reading are nowhere better illustrated in the narrative than during a visit to the Chicago Art Institute. While Jack sits and contemplates one of his favorite paintings, Pierre-Auguste Renoir's *On the Terrace*, for three hours, Pitsémine races through the museum and "sees everything there was to see" (74). It's possible to interpret this scene in several interesting ways. While gazing upon the Renoir, Jack enters into a blissful, trance-like state that suggests the power of art to be cathartic or transcendent. However, it also diminishes Jack's awareness of the world around him, a theme which Poulin will return to later, when Jack admits that he worries that writing can be a way of not living (98). By contrast, Pitsémine's hunger to see as many of the artworks as possible, as quickly as possible, is equally problematic. While her observation that books must be understood in relation to other books is an important insight, Poulin here seems to qualify that assertion by suggesting that it is also important to take the time to contemplate and reflect upon the various texts, in order to better understand them.

And it is not only the activities of travel and reading that are closely linked in Poulin's novel but writing as well. As Adams has pointed out, Jack Waterman's "creative future [as a novelist] depends on the ability to travel beyond national borders, coming to know, rather than insulating himself from, the American continent of which he is a part" (174).[8] Jack's romantic ideal of the writer, who writes effortlessly in a sudden burst of inspiration during which perfect sentence after perfect sentence come to him until he finishes a masterpiece and then passes out, is countered by Jack's decidedly uninspired reality as a writer:

> Jack Waterman was not very pleased with himself as a writer. He didn't like himself very much in general (he thought he was too thin and too old and too withdrawn), but what he hated about himself more than anything was his way of working. He had always had an image of the ideal writer, and he was utterly unlike that model. He ranked himself with those whom he called "the industrious class". patient and persistent but berett of inspiration or even impulses, he went to work at the same time every day and thanks to methodical and dogged work habits, managed to turn out his daily page. (31)

As this passage makes clear, Jack's romantic ideal of the writer is unrealistic and unattainable. Indeed, at one point, he tells La Grande Sauterelle that all his books are failures because they "don't change the world" (98). And just as Jack struggles with the disparity between his idealized image of his profession and his lived experience, so, too, does he desperately try to hold on to an idealized conceptualization of American history, one filled with heroes and heroic deeds. The statement, "And for Jack, in his heart of hearts, it was as if all the heroes of his past were still heroes," immediately invokes the epigraph to Poulin's novel: "We are no longer the heroes of history." With this epigraph, Poulin signals that one of his goals is to challenge romanticized notions of North American history, like those held by Jack. Instead, Poulin wants to shift the reader's focus to those who are left out of most historical accounts (the "ordinary people" like the pioneers who crossed the Oregon Trail) or those who have been wrongly misrepresented in many historical accounts, like the American Indians. Moreover, Jack's romanticized views are continually challenged by those of Pitsémine, who takes pains to carefully document the many massacres and atrocities inflicted upon Native Americans during the centuries of European colonization and settlement. Thus, in this novel, we have two competing visions of America. The first is neatly articulated by Jack in the following passage:

> He thought that in human history, the discovery of America had been the realization of an old dream. Historians said that the discoverers were looking for spices, gold, a passage to China, but Jack didn't believe a word of it. He maintained that since the beginning of the world, people were unhappy because they could not recover paradise on earth. They had retained the image of an ideal land and they searched for it everywhere. And when they found America, it was their old dream come true, and they would be free and happy. They were going to avoid the errors of the past. They were all going to make a fresh start. In time, the Great Dream of America had been shattered like all other dreams, but from time to time it was revived, like a fire smouldering under the ashes. It had happened during the nineteenth century when people traveled West. And sometimes, travelers crossing America found traces of the old dream scattered here and there, in museums, in grottoes and canyons, in national parks like the ones at Yellowstone and Yosemite, in the deserts and on the beaches like those in California and Oregon. (71)

It's significant that the latter part of the passage acknowledges that it is possible to find traces of the past in the present, but the traces that Jack and Pitsémine find during the course of their journey across the American continent are not evidence of a dream fulfilled. Instead, they find maps documenting genocide, gatling guns used to massacre American Indians, and so on. As Stephanie Athey puts it, Pitsémine is everywhere along the course of their journey confronted with evidence of her disinheritance.[9] The traces of the past they encounter seem instead to verify a contrasting vision of America, again articulated by Jack: "It's America. You start to read the history of America and there's violence everywhere. It's as if America was built on violence" (93). Even Jack's beloved Volkswagen bears traces of its manufacturer's own dark history: on the visor, written in German, is philosopher Martin Heidegger's assertion that "Language is the house of being," (he would go on to say that "In its house man dwells"). While an apt commentary on Poulin's insistence that our realities are inextricably bound up in language, as well as a metaphorical description of the Volkswagen bus, a traveling bookmobile within which our travelers dwell, it's also a reminder that Heidegger was a member of the Nazi Party and that the Volkswagen automobile began as a project authorized by Adolf Hitler and his Nazi regime. Thus, the mythology and rhetoric of a master race, and the genocide associated with this "trace of the past," mirrors that of the nineteenth-century United States' Manifest Destiny.

While Jack struggles to let go of his mythology of a heroic American past, Pitsémine also struggles to reconcile problematic aspects of Native American cultures with her beliefs. As Guenther puts it, "While La Grande Sauterelle's main function is often to counter the Eurocentric accounts of North American history, her displacements across the continent lead her to confront aspects of the First Nations cultures which resist idealization as well" (67). The first of these moments comes early in the novel when she spends a night sleeping beside the grave of the Mohawk Chief Thayendanegea (Joseph Brant) in the hopes of becoming reconciled with her conflicted identity. As she settles in for the night, she and Jack note that the grave of the chief's wife has no name on its marker. The following morning, disappointed that she feels no different, she tells Jack:

> First of all there's the business about Thayendanegea's wife who . . . I mean, she has no name and I spent part of the night wondering why, and I asked myself all sorts of questions about the old chief—how he treated his wife and all that—and then I wondered why he liked war, why he fought against the French and against the Mohicans from the Hudson River Valley, and finally, when I ran out of questions, I realized that I'd lost confidence in him. (60)

Here, for the first time, Pitsémine acknowledges the historical reality of intertribal warfare among Native American peoples. Significantly, she responds to this aspect of Native American history by recalling stories that her mother told her about her childhood that depict life in that earlier era in idyllic terms:

> In the fall she and her parents would leave the river bank where they'd spent the summer fishing and go north for the winter to hunt and trap, like the other families in the tribe. The women prepared the pelts that the hunters brought back to the cabins from their expeditions, and sometimes they'd take part in the hunt. The children were free to do whatever they wanted, and they weren't punished when they made errors in judgement. My mother said that the Indians never suffered from the cold and they never ran out of food. If a snowstorm went on for

> days and the hunters couldn't bring back fresh meat, they would eat the dried or smoked fish they'd brought from the south, and no one complained. (61)

Here, we see clearly how a nostalgic, idealized vision of the past serves as a kind of refuge from the mounting evidence that "America was built on violence," including inter-tribal conflicts. As the journey continues, Pitsémine takes Jack to Starved Rock, the site where, according to legend, the Illinois were exterminated by other Indian groups because they refused to join the Ottawa's intertribal alliance to fight against white settlers. Still later, toward the end of their journey, Pitsémine recounts in graphic detail the massacres of American Indians by the US military at Sand Creek, Washita, and Wounded Knee. She concludes this horrific recounting with the following observation:

> She simply wanted to say that the Indians, too, had been guilty of massacres. And all that had started in 1849, when the discovery of gold in California had brought to the Oregon Trail hundreds of thousands of adventurers of all sorts who, unlike the first emigrants, had no respect for anyone or anything, who destroyed everything along their way and provoked the Indians' wrath. (156)

Here, Pitsémine can only acknowledge atrocities on the part of Native Americans by rationalizing that they were justified due to white provocation driven by greed and even fixing them to a specific event. This is clearly a problematic statement and, I think, illustrates well the difficulty of coming to terms with historical evidence that challenges one's most deeply held beliefs. Ironically, Poulin himself is "guilty" of this same tendency, as literary critic Heather MacFarlane has persuasively argued. In her compelling essay, "*Volkswagen Blues* Twenty-Five Years Later: Revisiting Poulin's Pitsémine," MacFarlane observes that, despite the novel's emphasis on Native American histories, Native Americans themselves are curiously absent from the novel's present, reinforcing the myth of the vanishing Indian. With the exception of La Grande Sauterelle's mother, our travelers encounter no other Native Americans during the course of their transcontinental journey, even though they are traversing regions home to large native communities. The Native Americans in Poulin's novel tend to populate the past, not the present. As MacFarlane puts it, "Without exception, the myth of the dead and dying Indian is alive and well in *Volkswagen Blues*; the novel ultimately relegates Indigeneity to the past and leaves Pitsémine no choice but to choose hybridity over her Innu identity" (11).[10]

LITERARY TRAVELERS

As mentioned earlier, on the latter part of their journey, our travelers meet a "vagabond"—an elderly man who resembles Hemingway, claims to have known Kerouac, and is reading a novel by Jack London. This episode is designed to extend Poulin's interest in rethinking histories to literary history. London, Hemingway, and Kerouac are all larger-than-life figures in American literary history—and all three were prolific travelers whose works celebrated travel, adventure, the out of doors, self-reliance, and independence. Each is also representative of settings, cultures, and events important to the narrative—Kerouac through his French Canadian origins and as author of *On the Road*, one of the novel's touchstones; Hemingway through his Chicago origins and expatriate life in Paris; and London through his San Francisco Bay origins, socialist leanings, and participation in the Klondike Gold Rush. Moreover, all three authors wrote about tramps and tramping. Thus, by linking his vagabond to London, Hemingway, and Kerouac, Poulin is situating his travel fiction within a particular literary genealogy but doing so in a way that he complicates through his writer-protagonist. That Hemingway is also one of Jack Waterman's favorite writers should come as no surprise at this point, for just as Waterman holds a romanticized view of American history, which his experiences and interactions over the course of his journey challenge and undermine, so, too, does he hold a romanticized view of American literary history, which this novel challenges through the character of Waterman himself.

Though he greatly admires the tradition of these rugged writer-travelers, Waterman is anything but. Instead, he is withdrawn, inhibited, a homebody. As he himself puts it, "I'm a champion, too. I'm a champion at waking up in the middle of night and finding my old slippers with my toes and going to the kitchen in total darkness and making myself a hot chocolate WITHOUT EVEN TURNING ON THE STOVE LIGHT" (164). As Adams puts it, Waterman is "an anti-Kerouac whose writing methods and outlook on life could not be further from the Beats" (173). Ironically, the American writer whom Jack encounters during his journey, Saul Bellow, represents a very different strand of American literature, one that Jack is completely unfamiliar with. As Anne Maria Miraglia notes, Bellow was born in Québec to Jewish emigres (52). That both he and Kerouac would go on to become important American literary figures supports Adams' argument that Poulin is advocating a broader Pan-American perspective on the part of Québecois writers.

Kerouac, of course, is the most famous of the Beat writers, several of whom turn up still later in the novel, this time in the form of a photograph from the book *Beat Angels*. Poulin's text reproduces the photo and the caption for it, which reads "Lawrence Ferlinghetti, Minette Le Blanc, Peter Le Blanc, unidentified man, Allen Ginsberg, Harold Norse, Jack Hirschman and Bob Kaufman, Café Trieste, Grant Avenue and Vallejo, San Francisco, 1977, Photo by Diana Church" (201). As Waterman looks at the photo in the book, he recognizes the "unidentified man" as his brother. Poulin is able to again comment upon American literary history, as Adams persuasively argues, by using Théo as a metaphor for the forgotten Francophone strands of that history: "Here, [Théo] is a testament to the depressing fate of the Québecois author, who is doomed to be the unidentified figure at a literary gathering where he is surrounded by celebrated U.S. American writers" (175).

The novel ends in San Francisco, where Jack finally finds Théo, who is suffering from "creeping paralysis" and who does not recognize him. Théo's doctor discourages Jack from any more contact with Théo, telling Jack that trying to bring back the past could aggravate Théo's condition. This presents an ironic paradox, as Poulin's novel is all about bringing the past, North America's past, back to life. If we take the doctor's statement as a commentary on the kind of historical recovery practiced by Jack and Pitsémine during the course of the novel, it seems to suggest that to immerse oneself in the past is to become progressively paralyzed and unable to function in the present. And, indeed, there is substantial evidence to support this interpretation. Over the course of the journey, as Jack encounters increasing evidence that neither his brother nor the historical figures whom he reveres were the heroic individuals he thought they were, he suffers prolonged bouts of depression that incapacitate him. Pitsémine, also, is profoundly saddened as she is confronted with a series of vivid reminders of the violent, tragic histories of the American Indian cultures. Paradoxically, however, the farewell between Jack and Pitsémine in the final chapter seems to suggest a more positive development from their historical encounters—as Pitsémine chooses to remain in San Francisco, a culturally diverse city which she hopes will enable her "to come to terms with her twofold heritage, to become reconciled with herself" (220), while Jack returns to Canada under the enlightening and watchful eyes of the "gods of the Indians and the other gods" in America. While it's quite possible that Jack will fall back into his introverted routine and Pitsémine may continue to struggle with her self-worth and identity—the ending has a hopeful quality. Poulin seems to be suggesting that—by confronting the darkest aspects of the past, including their own histories and that of America—Jack and Pitsémine have enabled themselves to move on with their lives in a more positive way. And, as readers accompanying them on their journey, we, too, have learned anew about the multi-faceted history of North America, although Poulin would caution us that we will never be able to ever arrive at an accurate and comprehensive accounting of it, for precisely because language is our house of being, our histories will always be “half real and half invented,” just like Théo (98).

EPILOGUE: LOST IN TRANSLATION

As a literary critic who does not read French, I have had to rely on the English-language translation of Poulin’s original French novel for my analysis. Working from a translation, no matter how well done, is always problematic because the act of translation always creates changes to the original text, no matter how subtle they may be (just as I have argued earlier, the act of recording history inevitably alters it). However, the French and English versions of Poulin’s novel, as Hyman has carefully documented, present us with three substantive differences that inadvertently (and very fittingly) provide yet another opportunity to think about what Poulin is attempting to do with this novel. The original French edition contains, as a frontispiece, a map of North America identifying the routes of the early French and French Canadian explorers, as well as the Oregon Trail, with locations marked in both French and English. This map is not included in the English-language edition of *Volkswagen Blues*. Also missing from the English-language edition is a short but very important moment during Jack’s reunion with Théo in San Francisco: “Théo eut un mouvement de recul et, d’une voix tremblante: –I don’t know you, dit-il. Pendant un long moment, Jack demeura interdit.” Translation: “Théo recoiled and, with a trembling voice: ‘I don’t know you,’ he said. For a long while, Jack was speechless.”[11]

Finally, the Grogan epigraph to the English edition of the novel—“We are no longer the heroes of history”—does not appear in the original French edition. In response to Hyman’s inquiries, Poulin made clear that he intended the Grogan text to appear in the original publication, and the missing exchange between Théo and Jack was apparently an omission missed by both author and translator during the copy-editing stages of the editorial process. The reason for the omission of the map remains a mystery.[12]

These three substantive differences between the French and English versions of the novel provide a perfect metaphor for historiography itself, one of its major concerns, as it vividly demonstrates how information vital to an interpretation of the history the text is presenting is often omitted or altered. The omitted exchange between Théo and Jack is especially suggestive: the assertion by Théo that he does not know Jack disappears from subsequent retellings in much the same way that one might repress an especially painful memory, or in the manner that an atrocity might be omitted or redacted in a colonial history. Indeed, the parallels to Pitsémine, who is the product of two cultures who refuse to recognize her, are striking at this moment—Jack experiences the same intense sadness and pain in the moment of Théo's inability or refusal to recognize him that she has lived with her entire life. And the omission from the translation of Poulin's text weakens this striking parallel. However, by employing the literary theory advocated by Pitsémine and attempting to understand *Volkswagen Blues* in relationship to other books—in this case, fittingly, a translation of the "same" book—we find the "traces of the past" that had previously been overlooked, forgotten, and lost.

NOTES

1. I want to thank Rachel Adams and Caroline Levander, who introduced me to *Volkswagen Blues* in their 2007 seminar on hemispheric American literature. Rachel's analysis of this novel, part of her larger study of pan-American literature, has been enormously helpful in thinking about this clever and charming travel fiction, as were the discussions with the other seminar participants about this book.

2. For a fascinating account of the intertwined histories of Chinese immigration to the United States and Chinese cuisine, including the important role of chop suey in American culture, see Andrew Coe's *Chop Suey: A Cultural History of Chinese Food in the United States* (2009). Lest anyone doubt the power of travel narratives, Coe notes that Bayard Taylor's popular travel book, *A Visit to India, China, and Japan*, published in 1855, fueled anti-Chinese prejudice in the United States for at least three decades (117).

3. Adams, for example, asserts that *Volkswagen Blues* "embraces . . . the wanderlust that characterized Kerouac's life and career" (172) and "pays tribute to Kerouac" (173).

4. Anne Marie Miraglia argues that *Volkswagen Blues* "absorbs and transforms" its two most prominent intertexts, Jack Kerouac's *On the Road* and Gregory Franzwa's *The Oregon Trail Revisited*, which, as Miraglia notes, is itself a rewriting of Parkman's *The Oregon Trail*. See Miraglia, 53.

5. Adam Weisman remarks that it is "with deliberate irony that a Métisse—herself the physical evidence of the mixing of the European and native populations in North America—advances the philosophy of literary . . . mixing that sanctions the intertextuality" of *Volkswagen Blues* (9).

6. Roger Hyman observes that "neither Jack nor Pitsémine understand that the country and western music they hear on the road represents a white takeover and recycling of a specifically Black African American blues idiom, nor does Pitsémine understand that this white blues is a musical form that has attempted to do to American Black culture what imposed print literacy has done to her own" (10).

7. Adams, 149–87.

8. A number of critics, including Adams, Guenther, and Weisman, have characterized Poulin's project in *Volkswagen Blues* as one that rejects an insulated Québecois perspective for a broader American one.

9. Conversation with the author.

10. I'm grateful to Mary Stoecklein for bringing this essay to my attention.
11. My thanks to my colleague Reg McGinnis for providing this translation.
12. See the third endnote to Hyman's article.

Chapter Five

Jasmine's Shadow World of Migration

Bharati Mukherjee's novel *Jasmine* (1989) begins with an epigraph taken from a book on chaos theory by James Gleick: "The new geometry mirrors a universe that is rough, not rounded, scabrous, not smooth. It is a geometry of the pitted, pocked, and broken up, the twisted, tangled, and intertwined." The quotation highlights Mukherjee's interest in exploring how cultures and nations are ever more interconnected, tangled, and intertwined, as well as the dramatic rise in migrants, refugees, and displaced people traveling from poorer parts of the world to wealthier countries, reshaping entire cultural landscapes.[1] As such, Mukherjee's novel makes an interesting counterpart to *The Sheltering Sky*. Most obviously, it reverses the trajectory of the travel in Bowles's novel, as it tracks the journey of Jyoti, an Indian woman, from her home country to the United States. Thus, *Jasmine* provides an opportunity to consider the motivations for such migration, as well as the risks involved, and to compare those with the motivations and risks for Bowles's American tourists. Not only does the comparison reveal great differences but also some surprising similarities, as well, about which more in a moment.

Chaos theory famously popularized the concept of the butterfly effect (the notion that a butterfly fluttering its wings could set in motion forces that create a tornado on the other side of the globe), and that idea is very much present in the novel as well, both thematically as it deals with questions of predestination versus free will (another key aspect of chaos theory is that events which appear random or chaotic can actually be traced back to an initial set of conditions, which parallels the idea posed by the novel that what appears to be free will can actually be the culmination of predetermined conditions) and also in the evolution of the protagonist, who comes to wonder if she is the powerful product of just such a series of seemingly random events: "Time will tell if I am a tornado, rubble-maker, arising from nowhere and disappearing into a cloud" (241).

Mukherjee's novel moves between past and present, juxtaposing scenes of Jyoti's coming of age in India with her life in the United States, including the small town of Baden, Iowa, the setting of the novel's present. The juxtaposition of India and Iowa allows Mukherjee to highlight some striking and surprising similarities between the two locales. Both the region of India where Jyoti comes of age (Hasnapur) and the part of Iowa where the adult Jyoti, calling herself Jane, has settled are traditional farming communities that find themselves beset by radical change. In the case of Jyoti's Indian village, there are religious and political conflicts between Hindus and Sikhs, resulting in frequent violent incidents. The Green Revolution is also underway, and new technologies are rapidly transforming Indian society. Baden, Iowa, is also the site of much conflict, in this case, triggered by hard economic times for the rural agricultural community. Much of the farmland is being reclaimed by banks and sold to speculators and developers. Suicide and murder rates are up. Cultural differences are also becoming more pronounced here, too, as immigrants and refugees from Southeast Asia and other parts of the world settle here. The attitudes of Anglo Americans toward these new residents from other parts of the developing world, however, change much less rapidly, as the following passage reveals: "Out there. I am not sure what Mother imagines. On the edge of the world, in flaming deserts, mangled jungles, squelchy swamps, missionaries save the needy. Out There, the darkness. But for me, for Du, In Here, safety. At least for now. Oh, the wonder! the wonder!" (21). This passage, coming early in the novel, describes Jyoti's reaction to a comment made by a well-meaning older Iowan (the mother of Jyoti's lover, Bud Ripplemeyer), who is raising money for a church relief program. The ironic echo of Conrad's Kurtz in the final lines ("The horror. The horror."), suggest that the rhetoric of the civilizing mission that Conrad so effectively criticized as an integral part of colonialism is still very much alive and well and shaping the perceptions of the first world about the former colonies it once and, in some cases, continues to exploit. The irony and the qualifying statement, "At least for now," suggest that the civilized/savage dichotomy so central to that strand of colonial discourse is, of course, a fiction. We will soon learn, as the novel continues, that the first world is every bit as savage, dangerous, and violent as it claims its third world counterpart is.[2] As Heidi Slettedahl Macpherson remarks, "Mukherjee allows these gaps and misrepresentations to be situated in the text, and to stand alongside more accurate representations of Jasmine's past. Thus, Mukherjee is not, as some critics have suggested, setting up a clear demarcation between the progressiveness of America and the backwardness of India, but she is exploring how *perceptions* of this binary opposition are an underlying factor of Jasmine's U.S. life" (72).

The fragmented, non-linear narrative structure also gradually peels back layers of "Plain Jane's" facade to reveal a person who has experienced great suffering. In the India sections of the novel, we follow Jyoti's coming of age and her many struggles. She is a strong-willed, intelligent girl, but her father and paternal grandmother oppose her education, despite the encouragement of her schoolteacher, Masterji. Only with her mother's intervention is Jyoti allowed to continue her studies. Her education is to be short-lived, however, as her teacher is murdered by a gang of conservative Sikhs called the Khalsa Lions, and then her father dies after being gored by a bull. Jyoti falls under the care of her brothers, who introduce her to Prakash, whom she marries at the age of thirteen. Prakash is a "modern" man who sees many traditional Indian customs as backward and limiting. Renaming his very young wife Jasmine, he encourages her to be independent and an equal partner. Prakash dreams of moving to the United States with her and opening a small business. This happy period of Jyoti's life is cut short as Prakash, too, is murdered by the gang members.

In her grief, Jyoti determines to travel to the United States to burn Prakash's things and then kill herself. With her brother's help, she purchases forged documents and sets out on her journey to the United States. In one of the novel's most interesting chapters, Mukherjee offers an intriguing perspective on such emigration without legal documents:

> There are national airlines flying the world that do not appear in any directory. There are charters who've lost their way and now just fly, improvising crews and destinations. They serve no food, no beverages. Their crews often look abused. There is a shadow world of aircraft permanently aloft that share airlanes and radio frequencies with Pan Am and British Air and Air India, portaging people who coexist with tourists and businessmen. But we are refugees and mercenaries and guest workers; you see us sleeping in airport lounges; you watch us unwrapping the last of our native foods, unrolling our prayer rugs, reading our holy books, taking out for the hundredth time an aerogram promising a job or space to sleep, a newspaper in our language, a photo of happier times, a passport, a visa, a *laissez passer*.
>
> We are the outcasts and deportees, strange pilgrims vising outlandish shrines, landing at the end of tarmacs, ferried in old army trucks where we are roughly handled and taken to roped-off corners of waiting rooms where surly, barely wakened customs guards await their bribe. We are dressed in shreds of national costumes, out of season, the wilted plumage of intercontinental vagabondage. We ask only one thing: to be allowed to land, to pass through, to continue. We sneak a look at the big departure board, the one the tourists use. our cities are there, too, our destinations are so close! But not yet, not so directly. We must sneak in, land by night in little used strips. For us, back behind the rope in the corner of the waiting room there is only a slate and someone who remembers to write in chalk, delayed or to be announced or out of service. We take another of our precious dollars or Swiss francs and give it to a trustworthy looking boy

> and say, "Bring me tea, an orange, bread." . . . The zigzag route is the straightest. (100–1)

In this passage, Mukherjee effectively conveys the striking contrast between the privileged tourist's travel world and the undocumented refugee's shadowy counterpart and goes on to point out the bizarre moments when those two worlds intersect: "I sit with missionaries, with the deported, with Australian students who've fallen somehow into the same loop of desperation, which is for them adventure" (102). About this passage, MacPherson perceptively observes that Mukherjee is juxtaposing two very different types of travelers—those who travel out of necessity (and at great personal risk) for their survival, and those, like the Australian students, for whom travel is a luxury and an adventure.[3] And, as Robyn Warhol has pointed out, the passage is also noteworthy for the shift from a first-person singular to a first-person plural narration. In doing so, Mukherjee takes pains to underscore Jyoti's risky and difficult journey from the developing world to the developed one as representative of a much larger phenomenon. As Warhol puts it, "Her immigration is not singular, individualist; similar experiences continue to happen to countless people all the time. During her passage from India the narrator has no name at all, losing her specifically Indian identity and becoming part of a first-person plural that includes people in transit from everywhere in the world" (49), or what Warhol calls "the multicultural subject" (55). In this passage and in others throughout the novel, Mukherjee, in essence, documents the plight of the undocumented immigrant and legitimizes the status of those immigrants whom the US government would categorize as illegal.

Jyoti eventually reaches the United States by way of a ship, piloted by a Vietnam war veteran, who smuggles undocumented immigrants for a living. The vet, called Half-Face because of a scar on one side of his face, quickly recognizes that Jyoti is without resources or contacts in the United States and convinces her to spend the night in his motel room. Once there, he orders her to have sex with him, and when she refuses, he rapes her. Afterward, Jyoti enters the bathroom with the intention of killing herself but instead determines to turn the tables on her attacker. Slicing her tongue with a small knife given to her by one of the other refugees on the ship, she imagines herself as Kali, the destroyer goddess. Dripping blood from her mouth (as Kali is often depicted), she confronts Half-Face and slices his jugular, killing him. Careful to wipe down the room to erase her fingerprints, she exits and burns the remnants of Prakash's belongings and then ventures forth, "traveling light," the advice Half-Face had given her moments ago, and personifying her name, Jyoti, which means "light." Her use of the phrase, echoing the words of the man who raped her, suggests she feels no remorse for having killed him but also signals that she is releasing herself from a sense of old-world duty—now she is truly traveling light, making decisions for herself, not doing what tradition dictates or what others expect her to do.

The episode just described strengthens the ways in which Mukherjee's novel can be seen as an interesting counterpart to *The Sheltering Sky*. Like Kit, Jyoti ventures forth on her own, following the death of her husband, into an alien world, and, like Kit, her journey is cut short by a rape. But whereas Kit is so traumatized by the experience she devolves into a primitive being, Jyoti, as we will see, is able to emerge from the attack a much stronger person. Both women become omens—Kit becomes a cautionary tale, but Jyoti becomes the embodiment of Kali, just one of many identities which she assumes over the course of the novel.

Following her escape from the motel, Jyoti is rescued by Lillian Gordon, an American woman who offers sanctuary to undocumented immigrants from around the world. Lillian nurses Jyoti back to health and teaches her to walk and talk like an American, observing that most Americans can't imagine being from anywhere else (134–35). Through Lillian, Mukherjee adds another layer to her thoughtful presentation of immigration to the United States, as Lillian is eventually arrested and sent to prison for harboring undocumented workers. Having depicted the circuitous and dangerous route these immigrants must travel to reach the United States and the risks involved, such as the sexual violence and bodily harm posed by human traffickers like Half-Face, Mukherjee also depicts the risks involved for those who try to help refugees who come to the United States. Jyoti considers Lillian "an angel" and mourns her death in prison.

One of the most fascinating and valuable aspects of this novel is its examination of how immigration is reshaping the cultural landscape of the United States. Jyoti recounts the story of a retired Iowa farmer who suffers a heart attack in a California restaurant but cannot get assistance because all the undocumented workers flee, fearing the police. A similar scene unfolds in a news report Jyoti watches on television. In Iowa, Hmong refugees have settled in small farming communities and make quilts for Lutheran charities, and many of the doctors working in the hospitals there are from India and other parts of the developing world. Upon arriving in the United States, Jyoti marvels that she has yet to see what she considers to be an American face and has a similar reaction upon her arrival in New York City when she is picked up by an Afghan taxi driver. New York, she observes, is "an archipelago seething with aliens" (140). As Deepika Bahri remarks, "Each new immigrant who enters its [America's] space, legally or otherwise, shifts and transforms its boundaries. America is presented as a corporeal body undergoing constant change" (144). Moreover, as Kristin Carter-Sanborn notes, these episodes and others underscore the durability and power of stereotypes that Americans and Indians have of each other.[4] Bud, for example, is attracted to Jyoti precisely because she represents the exotic Orient ("Bud courts me because I am alien. I am darkness, mystery, inscrutability. The East plugs me into instant vitality and wisdom. I rejuvenate him simply by being who I am [200]"), just as Jyoti is completely disillusioned by the United States she encounters, one that does not meet her expectations: "In America, nothing lasts. I can say that now and it doesn't shock me, but I think it was the hardest lesson of all for me to learn" (181).

The next period in Jyoti's life deepens the novel's reflection upon immigration. She is taken in by Prakash's former professor, whom she calls Professorji. Professorji lives in a Queens apartment building with his wife and parents, and the building itself is nestled in a part of the city populated by a large immigrant community. This expatriate Indian enclave has the effect of retarding the assimilation of the immigrants to the American world outside of it. Within the community, within the apartment building, indeed within the apartment itself, traditional Indian customs and ways of life are carefully preserved:

> There were thirty-two Indian families in our building of fifty apartments, so specialized as to language, religion, caste and profession that we did not need to fraternize with anyone but other educated Punjabi speaking Hindu Jats. There were six families more of or less like Professorji's (plus Punjabi speaking Sikh families who seemed friendly in the elevator and politically tame, though we didn't mingle), and three of the families also had aged parents living in. (146)

Jyoti begins to chafe in this recreation of the traditional India that she had sacrificed so much to leave behind and comments upon the irony of the simulacrum: "Flushing was a neighborhood in Jullundhar. I was spiraling into depression behind the fortress of Punjabiness" (148).

What follows is one of the novel's most interesting episodes. After Professorji's father injures himself in a bathroom fall, Jyoti attempts to contact the Professor by calling all the nearby colleges and universities. When that fails, the Professor's wife manages to find a street address, which is to be used only in case of emergency. Jyoti describes what happens next:

> Flushing was not the downtown of dreams I'd conjured from the aerogram back in Jullundhar. And Professorji was not a professor. He was an importer and sorter of human hair. The hair came in great bundles from middlemen in villages as small as Hasnapur all over India. The middlemen shipped the hair in switches. Every weekday Professorji sat from eight o'clock till six on a kitchen ladder stool in a room he rented in the basement of the Khyber Bar BQ measuring and labeling the length and thickness of each separate hair. (151)

The hair, we learn, is used in wigs and scientific instruments, and the hair of Indian village women is especially prized for its natural purity. Jyoti marvels at the revelation: "A hair from some peasant's head in Hasnapur could travel across oceans and save an American meteorologist's reputation. Nothing was rooted anymore. Everything was in motion" (152). Ultimately, Jyoti contracts with the Professor to sell her hair as a means to purchase a forged green card. She departs without notice shortly thereafter.

This episode is remarkable in several ways. It again presents a shadow world of labor performed by immigrants to the United States. The former professor sorting hair in the basement for use in wigs and scientific instruments becomes the perfect metaphor for the global economy. If you live in or come from the developing world, chances are you provide either cheap labor or the raw material itself. In order to come to the United States, the professor has had to give up his status and his profession. The scene also echoes the novel's epigraph in the idea that seemingly random events are actually connected and can be traced back to small catalysts in such far-off places as India. As Jyoti puts it, "everything is in motion," including the world's population, and such movement, increasingly from the developing world to the more technologically-advanced nations, is greatly reshaping the former seats of empire. Furthermore, this section of the novel contrasts those immigrants like the professor, who seek to recreate their homeland in a new setting, with those like Jyoti, who seek to recreate themselves.

Following her stay with Professorji and his family, Jyoti secures a job as a nanny with Taylor and Wylie Hayes, Columbia University professors. This new chapter in Jyoti's life is again marked by a new name, Jase, given to her by Taylor, and is a relatively happy period of her life. Encouraged by Taylor and Wylie, who treat her as an equal, Jyoti becomes more independent. As she puts it, Jase "lives for today" and is a "prowling adventuress." She bonds with the Hayes's adopted daughter, Duff, and enjoys her new role as Duff's "day mummy." And she indulges in that most American of activities, consumerism. She becomes a compulsive shopper, buying items advertised on TV infomercials, until Taylor tells her that she can write "return to sender" on the package and send it back—an appropriate metaphor for the attitudes of some xenophobic Americans toward immigrants like Jyoti. Eventually, Taylor and Wylie separate, and Taylor and Jyoti fall in love, but just as they are on the verge of forming a new family together, Jyoti discovers that the man who murdered Prakash has tracked her to New York and is shadowing her movements. Terrified that he might harm her or her new family, Jyoti flees, impulsively choosing to go to Iowa because that is where the Hayes adopted Duff.

In the small town of Baden, Iowa, Jyoti finds work as a bank teller and soon enters a relationship with Bud Ripplemeyer, the bank's owner. He is fifty, and she is twenty-one. Though they don't marry, their relationship appears to be a committed one. Jyoti adapts surprisingly well to life in Iowa. She herself notes the similarities between her childhood home of Hasnapur and Baden: "The farmers around here are like the farmers I grew up with. Modest people, never boastful, tactful and courtly in their way. A farmer is dependent on too many things outside his control; it makes for modesty. They're hemmed in by etiquette" (11). Her traditional upbringing also allows her to relate well to the older generation of Iowans, who value large families and believe in hoarding necessities in preparation for hard times. Despite these points of connection, though, Jyoti is viewed by most of Baden's residents as something exotic, even though she now calls herself Jane Ripplemeyer.

About eighteen months after their relationship begins, Bud and Jyoti adopt a fourteen-year-old Vietnamese refugee named Du. Du is one of the most fascinating characters in Mukherjee's novel, and he provides her with another opportunity to explore a particular type of traveler, the displaced refugee. Du is a survivor of a horrific refugee camp, where he sometimes had to eat worms and insects for lack of food and where his mother and older brother were killed by "a jealous madman, after they'd gotten their visas" (18). Like Jyoti, Du appears to adapt quite quickly to life in Iowa. His school friends nickname him Yogi, and like Jyoti's late husband, Prakash, he has an affinity for things electronic and mechanical. Though he harbors a crush on his adopted mother, who is only seven years his senior, Du's relationship with Bud is cold and formal. During his days in the refugee camp, he learned to scavenge and adapt as survival strategies, skills that he continues to practice after moving to the United States. Unbeknownst to his adopted parents, he opens a bank account and saves a good-sized sum. He also tracks down his older sister, his only surviving family member, and secretly makes plans to join her in California, informing Jyoti only a few days before leaving. As these secrets come to the surface, Jyoti reflects on their meaning and on the different ways in which they each adapted to their new country:

> I am amazed, and a little proud that Du had made a life for himself among the Vietnamese in Baden and I hadn't had a clue. Aside from my Dr. Jaswani and from Dr. Patel in Infertility, I haven't spoken to an Indian since my months in Flushing. My transformation has been genetic; Du's was hyphenated. We were so full of wonder at how fast he became American, but he's a hybrid, like the fantasy appliances he wants to build. His high school paper did a story on him titled: "Du (Yogi) Ripplemeyer, a Vietnamese-American." (222)

In suggesting that her transformation has been genetic, Jyoti perhaps is trying to allude to an evolving, gradual, natural transformation. If we think of genetics as a sort of biological blueprint, it also is a nice way to offer another take on the novel's consideration of fate and destiny. Contrasted with this evolving persona is Du, the hyphenated immigrant, who has adapted to life in the United States but has not given up his connection to his native Vietnamese culture. Critics who dismiss *Jasmine* as an assimilationist novel have not accounted sufficiently for Du and the kind of immigrant that he represents.[5] As Carter-Sanborn astutely notes, Du vividly demonstrates Gloria Anzaldúa's "mestiza consciousness"—a tolerance for contradictions and ambiguity that is developed by having to exist in a liminal space.[6]

Jyoti also understands his carefully planned and calculated decision to leave her and Bud: "His eyes are glittery with a higher mission. Abandonment, guilt, betrayal: the boy in front of me would consider them banal dilemmas" (221). His abrupt departure, I think, becomes a model for Jyoti, who also decides to leave without telling Bud. Significantly, it is only to Du that Jyoti confesses that she killed a man—both understand that one must keep secrets to survive in this new environment.

One of the novel's great ironies is that its protagonist is disillusioned with the fluid environment she finds herself in, even as she celebrates her own series of transformations. Of her adopted country, Jyoti remarks, "It is by now only a passing wave of nausea, this response to the speed of transformation, the fluidity of American character and the American landscape" (138). She goes on to say, "In America nothing lasts . . . it was the hardest lesson to learn. We arrive so eager to learn, to adjust, to participate, only to find the monuments are plastic, agreements are annulled. Nothing is forever, nothing is so terrible, so wonderful that it won't disintegrate" (181). Yet, the evolution and transformation of Jyoti are indexed by the various names she is given or assumes, and each identity signals the beginning of a new chapter in her life. These transformations are often closely associated with various tragedies or violent events—when she is born, her mother tries to strangle her; Prakash names her Jasmine but then is murdered; Jazzy emerges out of the trauma with Half-Face; and Jane is a response to the reappearance of her husband's murderer—as Jyoti herself remarks, "There are no harmless, compassionate ways to remake oneself. We murder who we were so we can rebirth ourselves in the images of dreams" (29).

I'd like to conclude with a consideration of my favorite moment of the novel, which has to do with an unfinished game of *Scrabble*. After Du's departure, Jyoti enters his room and finds the game board with the words "deliquesce" and "scabrous," of which she remarks, "They are clues, but to what?" (225), and I think Mukherjee intends for the reader to take that comment literally. Scabrous is perhaps the easier of the two to make sense of, as the word appears in both the epigraph and again on the novel's first page, used to describe the protagonist. The word itself means "rough to the touch, difficult to handle tactfully, indelicate, off-color, salacious." Thus, the adjective links the new, interconnected, chaotic world of the epigraph with our ever-changing, traveling emigrant—driven out into the world by those rapid changes and becoming herself an agent of change.

Deliquesce is also powerfully suggestive in the context of the novel. The word means "to melt away or disappear as if by melting; to become liquid by absorbing moisture from the air (chemistry); to branch out into numerous sub-divisions that lack a main axis (botany)." Thus, deliquesce becomes a perfect description of the America Jyoti encounters: a place that attempts to absorb all difference and, by doing so, has become too dispersed and too transient, lacking any cohesive axis. Drawing on the work of William Bevis, Matt Burkhart has suggested thinking about "liquidity" in the capitalist sense: "through which capitalist ideology dissolves, absorbs, and exploits cultural distinctions" (19). But deliquesce may also be a warning for the multiple transformations that Jyoti undergoes—perhaps a way to suggest that taking on too many identities can cause a self to vanish. Finally, it may be a way to describe the survival strategies of the refugee and illegal immigrant, who survive by melting into the surroundings.

CONCLUDING REMARKS

As I've tried to show in this chapter, I believe that *Jasmine* is a much more complex novel than its detractors often acknowledge. One of the most common complaints about the book is that it re-presents to a Western audience a catalog of Indian stereotypes: poverty, illiteracy, sectarian violence, and the oppression of women, just to name a few.[7] Yet, Jyoti's struggle to overcome these problems is not accomplished on her own. She is aided by her mother, her teacher, her brothers, and her husband, each in their own way and often at great cost to themselves. This support system of Indian men and women of different generations suggests an interesting cross-spectrum of society that is looking for solutions to these problems and a way to resist the social, economic, political, and cultural forces that exacerbate them. Nor is the West, in this case the United States, presented as some idyllic refuge for the immigrant, or as a wonderful place of unfettered freedom and limitless opportunity. On the contrary, as Andrea Dlaska notes, the United States that Mukherjee depicts is every bit as troubled a place in its own way as India and every bit as beset by problematic and powerful social, economic, and cultural forces.[8] As noted earlier, transnational capitalism is having a tremendous impact in the United States, as well as in India: Mukherjee shows us the vulnerability of the undocumented workers laboring for low wages in the shadows, at the mercy of the police and INS; the professor who has legal residency in his new country but who has had to give up his profession because of a lack of economic opportunity and toils away in a basement sorting hair; and the small-time Iowa farmers who can't get a loan from the banks and whose foreclosed farms are then sold to wealthy corporations and developers. As Alison Graham-Bartolini remarks, "The sudden impersonality of the loan business is a concrete example of the changes taking place at both the social and economic levels in Baden [Iowa]" (100). And it's not just the undocumented immigrant like Jyoti who is vulnerable but even pillars of the community like Bud, who is shot and paralyzed because he becomes the focal point of the rage of a farmer who feels he's been cheated out of his land by the banks.

In terms of travel fictions, *Jasmine* is also a significant novel in its complicated and ambivalent consideration of immigration from third world to first. It carefully and vividly documents and dramatizes the risks involved and sacrifices made by those individuals who emigrate to the United States without papers, and it carefully and vividly reminds the reader of how much of the US economy is dependent upon those immigrants who labor without any legal protection. But at the same time, *Jasmine* does not pretend that mass migration to the United States is inconsequential. We see that it both enriches US society and causes disruptions, as well. One of the main points of the novel, it seems to me, is that, rather than take an inflexible or xenophobic stance toward immigrants, the US will need to find ways to adapt to this ongoing influx of peoples from around the world. Mukherjee herself has suggested that this was one of her goals in writing *Jasmine*, calling it "a meditation on whole peoples on the move and how America is or is not responding to this fact of de-Europeanisation."[9]

But, perhaps most remarkable of all is Mukherjee's break from one of the predominant themes in the literature of immigration—the importance of maintaining cultural traditions. Nowhere is this expressed better than at the end of *Jasmine*, as Jyoti decides to leave Bud to go west with Taylor and Duff. Of her decision, she claims that she is not choosing between men but rather between the promise of America and old-world dutifulness (240). Whereas many novels about the immigrant experience emphasize the importance of maintaining cultural traditions, Mukherjee celebrates change and the assertion of free will over some culturally predetermined fate. When traditional beliefs are used to oppress, Mukherjee suggests that a person has every right to break away from them, as Jyoti does. Nor does she believe one should set aside one's dreams because of what tradition dictates—those who do, destroy themselves, as Jyoti's Iowa neighbor Darrel does at the end of the novel. Darrel has inherited the family farm and struggles to find a way to make it profitable or risk losing the land and having it sold to developers who want to build a golf course. Though he continues to farm out of a sense of family loyalty and the dictates of tradition, his dream is to move to New Mexico and open a Radio Shack. Unable to pursue that dream or turn the farm around, he eventually kills himself, and his suicide can be interpreted as a warning for what happens when a person blindly follows the dictates of tradition, as the following passage makes clear:[10]

> When I was a child, born in a mud hut without water or electricity, the Green Revolution had just struck Punjab. Bicycles were giving way to scooters and to cars, radios to television. I was the last to be born to that kind of submission, that expectation of ignorance. When the old astrologer swatted me under a banyan tree, we were both acting out a final phase of a social order that had gone on untouched for thousands years.
>
> What I'm saying is, release Darrel from the land. There are different mysteries at work. Bud thinks it's a conflict between farming and golfing, but he's missing the point. Darrel is a romantic, just like him. The banker who steps out of marriage to live with an Indian is the same as the Iowan who dreams of New Mexico. They've been touched with the same virus. (229).

This passage juxtaposes the conflict between tradition and modernization in India with the same conflict in the United States. Fittingly, the novel ends with a reference to the tornado of the butterfly effect, as Jyoti wonders if she has become that destructive force as she moves in and out of people's lives in her adopted homeland. But Mukherjee's consideration of the impact of immigration from the developing world to the United States suggests that the answer to that question will depend on whether or not Jyoti's new homeland is willing to transform itself, as well.

NOTES

1. As Matt Burkhart puts it, *Jasmine* dramatizes how "diasporic movements from the so-called Third World are also causing substantial shifts in the cultural composition of occidental nations" (5).

2. And, as I hope my reading of this passage makes clear, I am not persuaded by Inderpal Grewal's assertion that *Jasmine* depicts a world in which "security and care are believed to be impossible in Punjab because of the inherent violence attributed to its populations, but in America safety, security, and 'ordinary' life are possible" (66–67). Quite the contrary: The America depicted in *Jasmine* is a volatile and dangerous place, particularly for undocumented migrants like Jyoti.

3. MacPherson, 64.

4. Kristin Carter-Sanborn, 577 and 581.

5. Grewal, for example, argues that *Jasmine* "suggests it is the responsibility of immigrants to become part of the American melting pot, rather than produce a separate, hyphenated identity" (67).

6. Carter-Sanborn, 581.

7. Ryan Paul Singh, for example, argues that "Mukherjee's narrative reinscribes colonial discourse by dealing in essentialized notions of the 'East' as oppressive and barbaric and the 'West' as the endpoint of culture and consumerist individualism as the high point of human development" (70).

8. Dlaska, 124.

9. As quoted in Dlaska, 125.

10. Alison Graham-Bartolini also makes this point in her essay on *Jasmine*.

Chapter Six

Small Place, Smallest Woman

If you read Jamaica Kincaid's book, *A Small Place*, this is what you will find. Though a mere eighty-one pages, it will confront you with many significant and difficult questions related to tourism and migration, using Kincaid's homeland, the island nation of Antigua, as a prime example of the perils of a small country dependent on the external revenue provided by tourism. You, the reader, will be firmly implicated in the exploitation of Antiguans by the tourist industry and made the target of the frustration, bitterness, and envy of those Antiguans who cannot afford to travel. And finally, you will be asked to think about tourism as a capitalist enterprise that mirrors and continues the exploitative relationships of the colonial period.

Published in 1988, *A Small Place* is a long essay neatly divided into three sections that focus, respectively, on tourism, Antigua's colonial history (it was a colony of England from 1632–1981), and post-colonial Antigua—followed by a brief epilogue. The fame of the book rests primarily with its opening critique of tourists and tourism, directly addressing a generic, white, middle-class American or European tourist visiting Antigua for the first time, in the style that I've mimicked in my opening paragraph. Kincaid deftly and quickly dispels the myth that tourism is an economic panacea for developing countries and goes on to delineate the many problems that the tourist industry creates in a small place like Antigua. She points out that the small island nation lacks the infrastructure and resources to cope with the vast numbers of tourists who visit each year, which, in turn, has a detrimental effect upon the island's environment and ecosystem. As Tereza Szeghi points out, one strategy Kincaid uses to make this point is to take classic images of tropical vacations and transform them into emblems of the detrimental impact tourism has upon both the native Antiguans and upon the island environment and ecosystem itself.[1]

> That water have you ever seen anything like it? Far out to the horizon, the colour of the sea is navy-blue; nearer, the water is the colour of the North American

> sky. From there to the shore, the water is pale, silvery, clear, so clear that you can see its pinkish white sand bottom. Oh what beauty! Oh, what beauty! . . . You must not wonder what exactly happened to the contents of your lavatory when you flushed it. You must not wonder where your bathwater went when you pulled out the stopper. You must not wonder what happened when you brushed your teeth. Oh, it might all end up in the water you are thinking of taking a swim in; the contents of your lavatory might, just might, graze gently against your ankle as you wade carefree in the water, for you see, in Antigua, there is no proper sewage-disposal system. But the Caribbean Sea is very big and the Atlantic Ocean is even bigger; it would amaze even you to know the number of black slaves this ocean has swallowed up. (13–14)

The linkage between tourism and slavery is clearly intentional. Kincaid insists upon understanding the tourist industry as a form of neo-colonialism that replicates the exploitative relationships of slavery during the colonial period. As Justin Edwards puts it, "For Kincaid, these [tourist] travelers perpetuate the slavery and imperialism practiced by their forefathers—not by keeping blacks in bondage chains but by contributing to an economy that keeps Antiguans shackled to tourism. Kincaid thus sees the citizens of Antigua, formerly British subjects, as still caught up in the hierarchies of colonization and economic imperialism. The island is, she suggests, unable to move beyond its status as a colonial nation" (77).

Kincaid also criticizes the tendency of the tourist to wrongly perceive local problems and poverty as something quaint or picturesque—reminding the reader that what is paradise for the tourist is often hell for the native. Related to this is the tourist's tendency to view the natives in a similar fashion, as backward, simpleminded, and unsophisticated. Such attitudes, in turn, breed tremendous resentment toward the tourist on the part of the local citizenry. And she dismisses the common argument that tourism is good because it provides jobs—the profits of the island casinos, for example, she notes, do not trickle down to the poor or working class (61). As Kincaid points out, the Antiguans who service the foreign tourists lack the economic resources to leave Antigua. They are forced to be either tourist providers or tourist attractions:

> That the native does not like the tourist is not hard to explain. For every native of every place is a potential tourist, and every tourist is a native of somewhere. Every native lives a life of crushing banality and boredom and desperation and depression, and every deed, good and bad, is an attempt to forget this. Every native would like a tour. But some natives–most natives in the world–cannot go anywhere. They are too poor. They are too poor to go anywhere. They are too poor to escape the reality of their lives; and they are too poor to live properly in the place where they live, which is the very place you, the tourist, want to

> go–so when the natives see you, the tourist, they envy you, they envy your ability to turn their own banality and boredom into a source of pleasure for yourself. (18–19)

She goes on to suggest that, over time, Antiguans even internalize this attitude, turning "the degradation and humiliation of their daily lives into their own tourist attraction" (69).

In response, Kincaid shrewdly reverses the tourist gaze that fixes the Antiguans as the quaint or exotic objects of the sightseeing tourist, pointing out to her reader that "the people who inhabit the place in which you [the tourist] have just paused cannot stand you . . . behind their closed doors they laugh at your strangeness" (17). In passages like these, Kincaid forces the reader to imagine themselves as spectacles put on display for voyeuristic consumption and to experience the discomfort of being objectified in that way.

She also does a skillful job of situating the tourist industry in Antigua in a larger context: "When you sit down to eat your delicious meal, it's better that you don't know that most of what you are eating came off a plane from Miami. And before it got on a plane in Miami, who knows where it came from? A good guess is that it came from a place like Antigua first, where it was grown dirt-cheap, went to Miami, and came back. There is a world of something in this, but I can't go into it right now" (14). That "world of something" is, of course, global capitalism, and the passage highlights the exploitation of cheap labor in developing nations to produce products that are then sold back to them at higher cost. Geographically, Antigua may be a small place, but its economy is inextricably linked with a global industry. And, like Dean MacCannell, Kincaid views tourism as a phenomenon driven by a kind of modern angst on the part of the privileged traveler, or, as she calls it, an "awful feeling of displacedness" (16).

In the book's second section, Kincaid reflects upon Antigua's past as an English colony and mourns the loss of ancestral culture and traditions that colonization wrought: "But what I see is the millions of people, of whom I am just one, made orphans: no motherland, no fatherland, no gods, no mounds of earth for holy ground, no excess of love which might lead to the things that an excess of love sometimes brings, and worst and most painful of all, no tongue. For isn't it odd that the only language I have in which to speak of this crime is the language of the criminal who committed the crime?" (31). As this passage suggests, Kincaid is also deeply troubled by the way in which her culture and traditions have become inextricably intertwined with those of her oppressors: "I met the world through England and if the world wanted to meet me, it would have to do so through England" (33).

At this point in the essay, Kincaid pivots, turns her attention to post-colonial Antigua, and asks herself if life there has improved for Antiguans since it became an independent, self-governing nation. Dejectedly concluding that it is not, she places the blame squarely upon a corrupt government and a populace which has been unable to reform or unseat it. She derides the post-colonial government for corruption, nepotism, kickbacks, state-sanctioned monopolies, and one-party rule. However, in what I find the most interesting and problematic section of the book, Kincaid also offers a harsh assessment of the average, working-class Antiguans living in the country that she has left behind but is revisiting as an expatriate. She accuses them of cultural myopia, of lacking any sort of historical perspective, and of becoming lost in victimization, all of which, Kincaid argues, results in a populace that is not able to distinguish between the trivial and the substantial. Among her more provocative statements are her claims that Antiguans cannot see themselves in a larger geopolitical or historical context; that they are unable to understand cause and effect (in thinking about how their actions might impact the future); and that this inability to think about their lives and actions within a broader context prevents them from solving the problems that bedevil contemporary Antigua:

> And might not knowing why they are the way they are, why they do the things they do, why they live the way they live, why the things that happened to them happened, lead these people to a different relationship with the world, a more demanding relationship, a relationship in which they are not victims all the time of every bad idea that flits across the mind of the world? And might not knowing why they are the way they are and why they do the things they do put in their proper place everyday and events, so that exceptional amounts of energy aren't expended on the trivial, while the substantial and the important are assembled (artfully) into a picture story ("He did this and then he did that")? (56–57)

Setting aside for the moment the clearly condescending tone of these remarks, one can parse them into those statements which are clearly false (that Antiguans have no sense of time or cause and effect) with those that may have some validity, for example, that people living in an insular place (nine miles wide by twelve miles long, as Kincaid notes) might tend to focus more on the local than the global. Yet, it's important to keep in mind that these remarks are made within the context of the question Kincaid poses as to whether self-ruled Antigua is a better place than colonial Antigua, governed by the English. Citing the corruption, nepotism, kickbacks, state-sanctioned monopolies, and one-party rule of post-colonial Antigua, Kincaid concludes that it is not, but earlier in the text, she points out that these problems cannot be disassociated from Antigua's colonial history: "Do you know why people like me are shy about being capitalists? Well, it's because we, for as long as we have known you, were capital, like bales of cotton and sacks of sugar, and you were the commanding cruel capitalists, and the memory of this is so strong, the experience so recent, that we can't quite bring ourselves to embrace this idea that you think so much of" (36–37). Yet, despite making this connection, Kincaid characterizes Antiguans as childlike, simpleminded, and naive—precisely the terms she criticizes tourists of using, as well as of the English before them: "Perhaps as you observe the debacle in which I now exist, the utter ruin that I say is my life, perhaps you are remembering that you had always felt people like me cannot run things, people like me will never grasp the idea of Gross National Product, people like me will never be able to take command of the things the most simpleminded among you can master, people like me will never understand the notion of rule by law, people like me cannot really think in abstractions, people like me cannot be objective, we make everything so personal" (36).

Again, this is precisely the same criticism that Kincaid makes of contemporary Antiguans. In addition to characterizing them as childlike, innocent, naive, and simpleminded, with no sense of time, she also accuses them of being lost in a sense of victimization. Relating an anecdote of two Antiguans who accidentally collide in a busy street and then carry on a long-running feud after the initial incident is over, Kincaid implies that the anecdote illustrates the lack of historical perspective of the entire culture and its inability to relinquish its status of colonial victimhood. While this latter idea is an interesting one to consider, namely that Antiguans are so obsessed with the abuses of the colonial past that it prevents them from solving the problems of the present, I find it ironic and troubling that Kincaid ascribes to contemporary Antiguans the same fraudulent qualities that she criticizes the English of attributing to them.

KINCAID THE EXPATRIATE

Kincaid's scathing critique of her fellow Antiguans is also problematic because, as she herself reminded us earlier in the essay, few Antiguans can afford to travel and leave the island—thus, if they do, in fact, lack a larger historical or geopolitical perspective (a thoroughly debatable point), the lack of economic resources must surely be taken into account. And while, unlike the foreign tourist, Kincaid is returning to the country of her birth and her childhood (she left when she was sixteen), she does have the economic means of the tourist to undertake this trip. Caryl Phillips states the problem succinctly: "The essay begs one question. Why does somebody who lives so comfortably outside think she has the right to criticize those who have to live inside? After all, it is easy to arrive with a return air ticket, make sweeping judgments and depart" (145). And, yet, while critics and scholars often acknowledge Kincaid's silence on the matter of her privileged status as problematic, the critical response too often has been a kind of feeble attempt to justify that silence. So, Phillips, for example, concludes that "Kincaid may reside in the United States but only somebody with her heart in Antigua could have written with such ferocity of purpose and self-revelatory hurt. Quite simply, she has a right to criticize because, irrespective of residence or nationality, she belongs" (146). Another critic, Emelia Ippolito, seems to accept Kincaid's representation of Antiguans as "naive" and suggests that Kincaid does so to provoke Antiguans to take action and "themselves expropriate power" (138). Unlike Phillips, however, Ippolito at least briefly offers a wider range of ways to think about Kincaid's position as expatriate:

> Nevertheless, in *A Small Place*, Jamaica Kincaid's departure from Antigua and her role as external observer who is at the same time an insider are never problematized. It is not clear how much of her anger is compensation for the privilege she gains from living in the United States, which played havoc with Antiguans. Unreserved in her condemnation, she might be using this attack to veil any privilege she gains from being a voluntary expatriate. While she wants people to assume responsibility for their actions, her departure from Antigua and her status as an expatriate are possibly acts and conditions she invites the population to consider embracing. She wants people to face up to the near surreal deformations Antigua has undergone. (138–39)

Other critics who have thoughtfully critiqued Kincaid's problematic stance in *A Small Place* include Alison Donnell, Barbara Edlmair, and Jeanne Ewert, who notes that Donnell calls Kincaid to account for an "'astonishingly arrogant propensity to speak for others,' which suggests the adoption of colonial discourse" (113). Ewert goes on to argue persuasively that, ironically, in Kincaid's travel writing, in particular *My Garden (Book)*, she depicts herself as a traveler whose behavior "is very little different from that of the tourists she criticizes in *A Small Place*" and adds "moments in *A Small Place* describing the appalling assumptions of rich American tourists about the lives of the people who inhabit the Caribbean, are repeated in her own travel narrative" (118). Also drawing on Donnell's work, Edlmair provides perhaps the most apt assessment of Kincaid's positioning (or lack thereof) in *A Small Place*:

> While this intermediate status, combining both the insider's and outsider's perspectives, does not disqualify her as a commentator on Antiguan history, it is nevertheless interesting that the ambiguity of her position is never problematized in the text. Indeed, Kincaid has been accused of deserting her birthplace and criticizing it from the safe distance of the affluent United States, a country with a clear political and economic interest in Antigua. Thus, there is an ambivalence present in the text concerning the voice of the narrator in *A*

> *Small Place*: on the one hand, the voice seems to claim an arrogant and absolute authority, adopting the colonialist discourse and pretending to be able to judge every one mentioned in the text; on the other hand, there is 'an indeterminacy of meaning, crucially linked to the undecidability about where this voice is speaking from and for whom.'[2] (79–80)

Another excellent discussion of *A Small Place* is provided by Antonia MacDonald-Smythe, who approaches the text through various post-colonial theories and finds some of Kincaid's discursive strategies problematic. She perceptively notes that Kincaid creates a panoptic eye that takes in both tourist and native, not unlike the imperial, all-encompassing gaze Mary Louise Pratt ascribes to earlier travel narratives:

> While the overarching interpretation of the tourist scene reflects the circumspective force of Kincaid's gaze, the power of her gaze has to be itself contextualized in the location of her observation post. Ultimately, the primary act of observation resides in Kincaid, the exiled writer who is at once insider and outsider, once native and now visitor to Antigua. However, the consequence of her inquisitive eye, that is, its potential reproduction of, rather than transformation of, the power systems at play in the tourist/native encounter needs to be foregrounded in so far as the perspective of Jamaica Kincaid, the multiply located witness, may be unconsciously repeating the biases of the powerful in her rendering of events. (147)

It is, she adds, "Kincaid alone who is allowed to tour" and who will "like the latter-day Columbus, publish her findings" (148). When directed at the tourist, MacDonald-Smythe continues, this approach inverts the tourist's presumed authority, but she also notes that Kincaid's perspective may unconsciously reinforce the bias of the powerful and wonders whether or not her speech is contaminated by the privileges of her various locations and whether her point of view is "calibrated with first world biases and new assumptions of superiority as she enjoys the privileges of her location as the 'darling' of *The New Yorker*" (152–53). Despite these reservations, MacDonald-Smythe suggests that Kincaid manipulates her ambivalent position as both native and tourist to reveal an oppositional consciousness and, further, that her refusal to locate herself is a deliberate act of resistance. *A Small Place* is, MacDonald-Smythe concludes, a critical intervention that opens new and promising possibilities for thinking about the West Indies and for attempts at cross-cultural representation, in terms that are not rooted in binary thinking or dichotomies (155). While I agree that *A Small Place* is indeed an important critical intervention, I remain troubled by Kincaid's condescending portrait of her fellow Antiguans and am not persuaded that her outrageous and demeaning characterization of them is, in fact, a deliberate rhetorical strategy designed to problematize her subject position without having to address it head-on.

EPILOGUE

As Ewert noted, eleven years after the publication of *A Small Place*, Kincaid published a book of essays about gardening and plant acquisitions called *My Garden (Book)*, which includes an account of a trip to China with other botanists entitled "Plant Hunting in China." Kincaid's response to the Chinese people she encounters and to Chinese culture is so mean-spirited (often provoked by the rigors of her trip, the difficulties of adjusting to cultural differences, etc.) it almost seems written to prove the claim she makes about tourism in *A Small Place*: "The thing you have always suspected about yourself the minute you become a tourist is true: A tourist is an ugly human being. You are not an ugly person all the time; you are not an ugly person ordinarily; you are not an ugly person day to day. From day to day you are a nice person" (14). Noting, as I do earlier, that Kincaid's reaction to Chinese culture seems at least in part exacerbated by the rigors of the journey, Ewert highlights yet one more irony when comparing the two texts: "She [Kincaid] has never suggested that American tourists in Antigua, like her exposed to a culture and living conditions much different than her own, might also suffer from brief lapses in sanity" (119).

I think what we can productively take away from Kincaid's ironic slip into the ugly tourist she criticizes in *A Small Place* during her China sojourn are two things. First, that transnational tourism has developed in such a way that it almost always involves, as Caren Kaplan puts it, "economic asymmetries" (62), no matter how well-intentioned the tourist may be, and, second, that travel from one place to another, even as a tourist, does not mean that the traveler or tourist cannot control his or her actions, thoughts, and reactions to the people and cultures he or she encounters. Simply put, admitting one is a tourist does not excuse bad behavior or exempt one from personal responsibility for how they treat others.

CANNIBALS IN THE METROPOLIS

If *A Small Place* provides a withering critique of tourism's impact upon developing nations, Clarice Lispector's brilliant short story "The Smallest Woman in the World" offers a similar dismantling of another type of travel fiction, the European narrative of exploration. These European narratives of exploration and discovery have proven to be so popular and durable that they continue to fascinate readers even today, and their protagonists have become legendary: Columbus, Cortés, John Smith, Cabeza de Vaca, and on and on to John Stephens and Percy Fawcett. One reason these stories were and continue to be so popular is that they are often cast in such a way as to reinforce the sense of cultural superiority of a European or American audience, as well as to provide a basis and justification for colonization. As Pratt observes, "Travel books . . . gave European reading publics a sense of ownership, entitlement, and familiarity with respect to the distant parts of the world that were being explored, invaded, invested in, and colonized. . . . They created a sense of curiosity, excitement, adventure, and even moral fervor about European expansionism" (3). Scholars like Pratt, Peter Hulme, and Rolena Adorno, just to name a few, have done an excellent job of critically examining such archetypal narratives of exploration. Lispector provides a similarly effective critique but goes about it in a very different way, using a fictional parody of such accounts to reveal the travel fictions that they are often based upon.

Lispector's story takes as its point of departure one of the quintessential moments of colonial travel narratives—the European explorer making first contact with a non-European culture. The story begins in equatorial Africa, where French explorer, "hunter and man of the world" Marcel Pretre hears rumors of a community of pygmies deep in the jungle (447). In Eastern Congo, he discovers the "smallest pygmies in the world" and among them, "the smallest woman in the world," whom, it turns out, happens to be pregnant (447). The story's opening is full of cues that Lispector is mindful of, and playing with, the conventions of these tales of colonial mastery that European readerships love to tell themselves. For example, here is how Lispector describes Pretre's reaction upon meeting the smallest human being in the world: "Feeling an immediate necessity for order and for giving names to what exists, he called her Little Flower. And in order to be able to classify her among the recognizable realities, he immediately began to collect facts about her" (448). Such behavior clearly marks Pretre as one of the "gentleman explorers" that emerged during the Enlightenment. As Pratt puts it, "In the second half of the eighteenth century, whether or not an expedition was primarily scientific or the traveler a scientist, natural history played a part in it. Specimen gathering, the building up of collections, the naming of new species, the recognition of known ones, became standard themes in travel and travel books" (27). Here, the specimen is the smallest woman herself. We are given her exact height—"seventeen and three quarter inches high"—suggesting that Pretre has measured her. He also names her Little Flower, never pausing to consider that she has a name of her own, and he describes her to the Press as "black as a monkey" (447). All these activities—the note-taking, the measuring, the fact collecting, the classification, the naming, the dehumanization—are part and parcel of colonial travel narratives. But after signaling the genre as her focus, Lispector is quick to put to rest any doubts about whether her story will engage critically with it—these actions of colonial mastery are immediately undermined by Little Flower's inappropriate responses. Her reaction to Pretre's declaration of her new name ("You are Little Flower.") is to unselfconsciously scratch herself "where no one scratches," unnerving the explorer and forcing him to look away (449). To paraphrase Trin-T-Minh-Ha, through her inappropriate response, Little Flower remains unappropriated, if only for a moment. Describing this scene as a "moment of crude burlesque," Claire Williams observes that Lispector's presentation of the episode has another important effect, that of highlighting the voyeuristic, objectifying gaze of the Western explorer: "Part of [Pretre's] shock is that *she* shows no shame, yet *he* feels it because he is looking voyeuristically at something which would be denoted as private and intimate in his culture" (172, emphasis in original). Thus, Williams argues, Lispector effectively parodies "the frequently romanticized ethnographic confrontation between cultures and, by implication, problematizes all instances of self/Other engagement" (172). Judith Rosenberg offers a similar interpretation, remarking that "Instead of conquest, Pretre experiences embarrassment, shame, and a loss of control that is intensified during the narrative in proportion to his diminishing comprehension" (71).

It's worth taking a moment to say something about the space in which this encounter takes place. The meeting between the smallest woman and the French explorer occurs in what Pratt calls a contact zone, which she defines as places where formerly separated cultures come into contact and begin to interact with one another. Often, these contact zones are marked by conflict and "asymmetrical power relationships" that historically have led to such oppressive institutions as slavery and colonization. Pratt points out that a phenomenon of the contact zone is something she calls transculturation—the exchange of influence each culture has upon the other. In other words, not only is the subordinate group changed by the dominant one, but also vice versa. Pratt emphasizes that the subordinate group is not powerless—it can, to a certain degree, modify, resist, or subvert the changes being imposed upon it by the more powerful group.[3] Thus, returning to Lispector's story, it is not surprising that the smallest woman reacts in a way that subverts Pretre's action—naming, after all, is but the first step toward taking possession of something.

Pratt adds, "While the imperial [metropolis] tends to imagine itself as determining the periphery (in the emanating glow of the civilizing mission or the cash flow of development, for example), it habitually blinds itself to the reverse dynamic, the powers colonies have over their 'mother' countries. For instance, empires create in the imperial center of power an obsessive need to present and re-present its peripheries and its others continually to itself. It becomes dependent on its others to know itself. Travel writing, among other institutions, is heavily organized in the service of that need" (4). This obsessive re-presentation of the Other and the dependence it engenders is also brilliantly captured and examined in Lispector's story. She abruptly shifts her focus from the periphery to the imperial metropolis, as the narrator tells the reader that a life-size color photograph of Little Flower is published in the Sunday newspapers of an unnamed metropolis (though there are clues to suggest it could be Rio de Janeiro or another urban center in Brazil). In this second third of the story, Lispector carefully tracks and details the different reactions to the smallest woman's photograph by different readers. One woman reacts with disgust, another feels a "perverse tenderness," while others feel sorrow and pity. Some view her as an animal and liken her to a dog or a monkey. A young boy, upon seeing the photo, wants to possess the smallest woman so that she can be his plaything, while, in another household, a teenage girl desires Little Flower so that she can become the family servant. In this sampling of reactions to the "discovery" of Little Flower, we see that the supposedly more civilized group is anything but. In their desire to possess her and in their unwillingness to see her as a fellow human being, the metropolitans reveal their own primitive, base natures. Lispector, in effect, takes one of the key tactics of colonization and turns it against the colonizers themselves, as she catalogs and inventories their reactions to Little Flower, just as Pretre catalogs and inventories Little Flower and her people.

Perhaps the most interesting reaction to the photo, however, comes from a young boy's mother, who remembers a childhood story told to her by a cook about life in an orphanage:

> The orphans had no dolls, and, with terrible maternity already throbbing in their hearts, the little girls had hidden the death of one of the children from the nun. They kept the body in a cupboard and when the nun went out they played with the dead child, giving her baths and things to eat, punishing her only to be able to kiss and console her. In the bathroom, the mother remembered this, and let fall her thoughtful hands, full of curlers. She considered the cruel necessity of loving. And she considered the malignity of our desire for happiness. She considered how ferociously we need to play. How many times we will kill for love. Then she looked at her clever child as if she were looking at a dangerous stranger. And she had a horror of her own soul that, more than her body, had engendered that being, adept at life and happiness. She looked at him attentively and with uncomfortable pride, that child who had already lost two front teeth, evolution evolving itself, teeth falling out to give place to those that could bite better. "I'm going to buy him a new suit," she decided looking at him, absorbed. Obstinately, she adorned her gap-toothed son with fine clothes; obstinately, she wanted him very clean, as if his cleanliness could emphasize a soothing superficiality, obstinately perfecting the polite side of beauty. Obstinately, drawing

> away from and drawing him away from, something that ought to be "black as a monkey." Then, looking in the bathroom mirror, the mother gave a deliberately refined and social smile, placing a distance of insuperable millenniums, between the abstract lines of her features and the crude face of Little Flower. But with years of practice, she knew this was going to be a Sunday on which she would have to hide from herself anxiety, dreams, and lost millenniums. (450–51)

The mother's reaction here brilliantly exposes the lie of so much colonial discourse about other cultures. Here, the mother's anxiety is rooted not in some perceived difference between herself and the smallest woman but rather in their shared humanity. Her sense of superiority is as superficial as her refined and practiced smile or the new suit she wants to buy her son. While she would like to think that she is superior to the smallest woman, she recognizes on some level that the smallest woman is no more or less savage or civilized than she or the people of her own culture. Little Flower's humanity and her perceived difference or otherness from those metropolitan subjects pose a challenge, one might even say a threat, to their sense of cultural superiority. Thus, they respond in ways which seek to shore up that sense of superiority: by viewing her as an animal, servant, plaything, possession. They deny her humanity. This ferocious need to separate themselves from their shared humanity with Little Flower is the transculturation—it is the effect she has upon them. By tracing all these different reactions to Little Flower's photograph in the newspaper, Lispector illustrates the horrific nature of colonialism itself: it is a mechanism to own, to enslave, to dehumanize. And these ends are concealed within different rhetorical registers: of pity, of the need to civilize, of savagery. These so-called civilized and sophisticated readers want to own Little Flower as a plaything, a servant, a curiosity. They sexualize her, primitivize her, and dehumanize her, and yet their responses reveal their own primitive and base natures and desires. And in Lispector's portrayal of these reactions, she vividly demonstrates how the very existence of Little Flower out in the "periphery" challenges those in the metropolis to try to come to terms with her and her culture. Earlier in the story, Lispector made a point to mention the Bahundes, a group of cannibals living in the Congo who prey upon Little Flower's people, and the invocation of this fearsome, mythic figure enables the reader to see that the cannibals exist in the metropolis as well: Although not devoured in a literal sense, Little Flower is avidly consumed by the reading public; we see that she has already been commodified for European-American consumption. Lispector skillfully plays with notions of civilization and savagery—forcing her readers to see the readers in the story as the uncivilized savages, not the Likoulas, and, as Williams remarks, to examine their own reactions to Little Flower: "By representing such a range of reactions, Lispector anticipates the response of her readers, thus weaving them into the narrative and forcing them to be conscious of the reading process" (178).

For the final third of the story, Lispector returns to Africa and reminds us that Little Flower is pregnant: "Methodically, the explorer studied that little belly of the smallest mature human being. It was at this moment that the explorer, for the first time since he had known her, instead of feeling curiosity, or exaltation, or victory, or the scientific spirit, felt sick. The smallest woman in the world was laughing. She was laughing, warm, warm—Little Flower was enjoying life" (452). Once again, Lispector parodies colonial narratives of exploration and ethnography in the objectification of the smallest woman and the continued efforts on the part of the explorer to examine her as some sort of scientific project. And, just as she undermined his earlier effort to name her with her "inappropriate action," here she completely unsettles him with that most human of actions: laughter. Pretre's discomfort with her laughter suggests that he finds it somehow incongruous that someone whom he perceives as such an oddity should be capable of the same human emotions as himself. Lispector reinforces this critique a few lines later: "She [Little Flower] loved that sallow explorer. If she could have talked and had told him that she loved him, he would have been puffed up with vanity. Vanity that would have collapsed when she added that she also loved the explorer's ring very much and the explorer's boots" (453). Here, the explorer's inflated sense of self-importance is undermined by Little Flower's thoughts and actions. From her point of view, he is every bit of an exotic curiosity as she is to him, and she reduces him to the same level as his boots or ring, providing a striking example of what Pratt calls "reciprocal vision," viewing the European subject through the eyes of the Other. As Williams puts it, "*Pequena Flor* [Little Flower] meets the ethnographic gaze of the spectating male and measures him on her own terms, curious and amused, looking him straight in the eye rather than lowering her gaze in deference and refusing to act as a mirror for the greatness he expects her to reflect onto him" (179). Moreover, Little Flower's fascination with Pretre's boots and ring suggests a materialism that most of those newspaper readers back in the metropolis would not expect of her. And, in lumping Pretre together with his ring and boots as exotic curiosities, we are reminded that she is no more immune from seeing him in dehumanized terms than those newspaper readers back in the metropolis responding to her image or the explorer himself.[4]

Gradually, the explorer composes himself, adjusting "his symbolic helmet," and resumes his note-taking. He then looks into Little Flower's eyes and believes they tell him, "It is good to own, good to own, good to own," and he winks in a moment of imagined collusion. This penultimate moment in the story is a complicated one. While it appears to reinforce the notion that materialism is a common human desire, the fact that it is articulated through the explorer's interpretation of what he believes he sees in her eyes calls it into question and reveals once again the colonial tendency to believe it can read its colonial subjects like a book, a tendency Lispector dismisses in the brilliant final paragraph of the story, which offers one last response to Little Flower's photograph: "'Well,' suddenly declared one old lady, folding up the newspaper decisively, 'well, as I always say: God knows what He's doing'" (454). Lispector leaves us with one more common response to radical otherness: Rather than try to understand Little Flower and create intellectual space for her, the old woman shrugs her shoulders and decides not to think about her at all.

To sum up, Lispector's short story is a brilliant demonstration of transculturation, the products of interaction between colonizer and native in the contact zone, an often uneven but still mutual exchange. It's easy for us as readers to imagine the disruptive potential that the French explorer's discovery will have upon Little Flower's society; and it is precisely for this reason that Lispector devotes so much of the story to detailing the various reactions to the discovery back in the metropolis. By doing so, Lispector enables her reader to see that a disruption occurs there as well: Little Flower's existence challenges the metropolitan residents to make room for her intellectually and they, in a variety of responses, are unable to do so, except in ways which would keep Little Flower under their control—as a servant, a plaything, an exotic curiosity, and so on, right down to the woman at the end of the story who would rather not have to think too much about a shared humanity with Little Flower. And it is this subconscious recognition of a shared humanity with her that is clearly the source of anxiety Little Flower provokes within these readers—anxiety that can only be dispelled by exoticizing or primitivizing her, but by using the technique of reciprocal vision in her narrative, Lispector reminds us that what is exotic or primitive depends upon who is doing the looking. The meeting between Little Flower and the explorer triggers a transcultural disruption that occurs not just in the contact zone but also radiates back to the metropolis, as well—as Lispector's travel fiction so brilliantly demonstrates.

NOTES

1. Conversation with the author.

2. The quote within the quote is taken from Alison Donnell's essay, "She Ties Her Tongue" (107).

3. See Pratt's *Imperial Eyes*, 4–6.

4. Reading Little Flower against the other protagonists of *Family Ties*, the short story collection where "Smallest Woman" first appeared, Marta Peixoto observes that

Little Flower is exceptional in her self-possession and "tranquil independence," calling her a "wry symbol of effective female power, though not of the means to attain it" (38).

Chapter Seven

A Land Made of Its Representations

In 1972, Lynne Tillman came to Amsterdam "intent on becoming a writer, and in the spirit of the Lost Generation and earlier American writers, believed that living in Europe, or out of America, as expatriate or alien, was what she needed to free herself or lose herself, and write" (*What Would Lynne Tillman Do?* 49). The year prior, she had read Paul Bowles's *The Sheltering Sky* and Jane Bowles's *Two Serious Ladies*, both of which were to influence her greatly, and in early 1972, she was given a rare, out-of-print 1932 anthology called *Americans Abroad*, featuring selections by such famous expatriate authors as Stein, Pound, and Eliot. The anthology gave her the idea to compile and edit a new, contemporary collection of writing by Americans abroad, which she pitched to the editor of a Dutch publishing house, who gave her a contract and an advance for the proposed volume.

One of the first writers she contacted was Paul Bowles (Jane had passed away in 1973). As Tillman puts it, "To me, he was the preeminent American abroad (the term is aptly dated), and I was determined to have him in the book" (38). To Tillman's delight, Bowles agreed to contribute. Thus began a decades-long correspondence and friendship that would last until his death in 1999: "We wrote about domestic life—collapsing roofs—and dreams we had. [Bowles] typed his letters on white, crinkly airmail paper. His signature, in black pen, was neat and without flourishes. I have a couple of letters on green airmail paper written entirely in his legible hand. In one he wrote: 'Place seems to have become unimportant'" (40). As one of his generation's preeminent travel writers, Bowles's remark is striking in its implicit criticism of the kind of cultural homogenization that has taken place since the Second World War. It is also a reminder of how important and central to his fiction so-called exotic places are and how essential travel abroad was to his literary production. The same is true for Tillman. In an interview with Peter Nicholls, she remarked that "I needed, with my insecurity about writing, just to be in another place. To survive. All the otherness around me allowed me to express my difference, my Americanness. It was easier to be an American in a way and to find my own language in the midst of people who weren't. The paradox is that you come closer to discovering what home is when you're far away from it" (270). She further observed that travel provides a useful perspective and a new way of thinking about the world, despite the difficulties associated with it: "I don't actually find travelling that enjoyable, but on the other hand I have a greater fear of stasis. I mean I have a real fear that if I sit in my apartment, for a very long time, I'll lose any kind of perspective I have, that I really won't be able to see my thoughts at all. They'll simply be the wallpaper everything else is and I'll just accept everything. My fear is that I'll just accept all the ways in which I'm limited because I won't any more see them as limits. You begin to recognize your limits when you're up against the unfamiliar" (271).

Sadly, the expatriate anthology was never published, and a subsequent attempt by Tillman to adapt *Two Serious Ladies* to film was also never realized due to a legal dispute over who held the rights to the novel. Nevertheless, her friendship and correspondence with Bowles would eventually lead Tillman to write her own novel about an American abroad, *Motion Sickness* (1991).

Like Kincaid's *A Small Place*, *Motion Sickness* is an astute critique of the privileged first-world traveler—although, this time, the setting shifts from the developing world to the major cities of Europe, the Near East, and North Africa. A plot summary of Tillman's novel is difficult because there isn't much of one. The book follows the travels of the unnamed narrator, a young American woman, as she enacts a kind of late-twentieth-century grand tour of the European continent. She does not follow a planned itinerary as a more conventional tourist might; instead, she drifts from place to place and city to city, often impulsively, and, as the novel progresses, she begins to shuttle back and forth between places she has already visited, so that her travels in Europe begin to form a kind of closed circuit.

The novel's first line, "PARIS: I am in my hotel room, on the bed, reading *The Portrait of a Lady*, and nursing an illness I might not have," is emblematic of its concerns: the artificial tourist bubble the first-world traveler occupies when abroad, the ways in which travel narratives of different kinds greatly influence the tourist's perception of the places they visit, and tourist angst—a nagging uncertainty that there is something awry. Additionally, the opening chapter's title, "Tourist Attractions," is a useful pun: Tillman's novel is also about the tightly knit expatriate community in Europe that well-to-do travelers inhabit, moving around and through a closed circuit of destinations, as though drawn to each other by some powerful, unseen force.[1]

Moreover, as Tillman ably demonstrates with this novel, it is not only the expatriate community in Europe that forms a kind of closed system but also the community of readers of travel narratives and travel fictions. *Motion Sickness* displays Tillman's deep familiarity with travel narratives and travel fictions at every turn. As the book's opening line suggests, reading is inextricably linked with traveling for Tillman's protagonist; the two activities occur in tandem and influence each other. *Motion Sickness*, like *Volkswagen Blues*, uses its intertextuality to critique both the genre and tradition it belongs to. It is no coincidence that Tillman's novel begins with a reference to James's *The Portrait of a Lady*. Like Isabel Archer, Tillman's protagonist is an American woman who has come to Europe following the death of her father: "I'm playing with the idea that I'm writing about an American who's traveling, and Henry James obviously wrote that book," Tillman observed in an interview. "Here you have one book in relation to another book, one heroine being constructed and opposing herself to this other heroine. And, of course, it's a kind of joke."[2] Tillman is being a bit disingenuous here. While it's certainly a playful nod to the knowing reader, the James allusion is part of a deliberate intertextual strategy that references a dizzying litany of other travel fictions, travel narratives, and literary travelers—including *The Sheltering Sky* and *Jasmine*. The point I wish to make here, and one that I think Tillman is making in her novel, is that the writers of travel fictions are often responding to other works of travel literature, as much as they are to other people, places, and cultures.

Both the form of *Motion Sickness* and its narration are designed to reflect its focus on travel and tourism. While the narrative is superficially structured around the narrator's movements from place to place, the narrative itself is a kind of free-associative stream of consciousness triggered by each new travel experience. In the book's second chapter, the narrator remarks that "all conversation leads from subject to subject, a horse that trots, canters, then gallops away from one's initial point of departure" (16), a perfect metaphor for the narrative constructed and employed by Tillman herself. This is a narrative that is almost entirely introspection and reflection upon the travel experiences described, and the constant flitting from idea to idea in the narrator's mind, as well as the frequent digressions in her recounting of events, is depicted here as very much the product of the travel experience—each encounter with the new triggers another cascade of associations. The unfamiliarity, disorienting, and isolating effects of travel upon the solitary traveler increase the act of introspection. For this narrator, movement through foreign spaces encourages flights of fancy and daydreams, often based on novels and movies, about which more will be discussed in a moment.

In one of these introspective moments, the narrator confesses, "My mobility embarrasses me" (186), highlighting one of the most prominent, recurring themes in *Motion Sickness*—the privileged position of the first-world traveler.

The narrator's admission underscores that the ease with which she can move about Europe on a moment's notice or whim clearly makes her uncomfortable and, I think, alludes to the novel's title—the anxiety the narrator feels about her privileged position as an American tourist abroad.

Though at different points the narrator frets about not having enough money to stay in Europe, we know that she is able to travel about quite comfortably on the money she accumulated in the United States and with financial support from her mother (59). Thus, if her money runs out, she can return to the States if she needs to, unlike Jyoti of *Jasmine*, for example. Tillman makes clear that an expatriate is something altogether different from an exile: her protagonist is keenly aware that her expatriate status is a privileged one, not to be conflated with that of other displaced people, whom she believes, unlike her, run the risk of being trapped in a "kind of limbo . . . a not-anywhere" (21). Of her own displacement, she remarks, "Voluntary and individual diaspora is a luxury" (21). She elaborates on this point later in the chapter:

> I feel out of place and know that I'm right to feel out of place. Travel unsettles the appropriate. You're bound to be inappropriate. Which is why I don't feel the intense embarrassment some do at not being able to speak foreign languages correctly. It seems to me that one of the privileges of travel is never to fit in. And not to fit in, not to be able to, is a kind of freedom. One of the freedoms that money can buy, like buying a hotel room in which one is psychologically unburdened and can act out guilty pleasures, capitalist ones, no doubt. (32)

As this passage makes clear, Tillman's narrator, unlike the immigrant or refugee, feels no pressure or need to assimilate into the various foreign cultures she moves through—and she correctly recognizes that lack of social pressure as a kind of privilege, a kind of freedom. The closing line of the passage reminds us that this privilege is very much a form of economic power, symbolized by the hotel room, which, for the first-world traveler, becomes a kind of fortress or sanctuary from which to view the foreign culture at a safe remove:

> The hotel patio overlooks the Socco, offering, as in the tourist brochure, a panoramic view of the city. With binoculars I could watch individual men opening their small stores, see their expressions, and follow women in the crowded market at the bottom of the road, and see the intensity of their lives at a distance, a safe distance. I once read that safety is an acceptable level of risk. The patio is always hard to leave. . . . This room is difficult to leave. I like it, with its window to the world. I like windows. Windows of opportunity. (120)

Like Kit in *The Sheltering Sky*, Tillman's narrator often gazes at the world from behind her hotel window, and like Kit, she is passive, and her gaze is that of both a tourist and a voyeur. The "opportunity" and "guilty capitalist pleasure" provided by her money, her economic power, is the privilege of consuming the foreign space as a paid spectacle, without fear of condemnation or repercussion.

The idea that travel provides an opportunity for voyeurism and that sightseeing, an activity closely associated with travel, is a type of voyeurism, is a recurring theme in *Motion Sickness*. The narrator's love of spy stories, both historical and fictional, for example, underscores her own predilection for spying on others. Sightseeing, for the narrator, involves people watching, and she is continually imagining narratives for those she spies upon. (Curiously, only occasionally is she aware of the fact that, as a foreigner, she is an exotic curiosity for the local people she moves among, as, for example, when she and two English brothers drive across the Italian countryside in an English taxicab.) This tendency is so pronounced that her friend, Jessica, accuses her of "being voyeuristically engaged in other people's lives" (76), a charge she only mildly disputes. Yet, on some level, the narrator seems to recognize that her fascination with other people has to do with their "otherness," their inscrutability. As she puts it, "Not knowing the language or the customs, to me everyone and everything is thoroughly clothed, wrapped in the thick mysterious outwear of the other. They have foreign tongues and foreign skins that protect them from outsiders. Or aliens" (76).

According to our narrator, travel unsettles the appropriate—as one travels, one feels out of place, and that sensation is to be expected. Yet, significantly, this sense of not belonging does not produce the embarrassment her mobility does. Instead, she finds it exhilarating—and she comes to view her dislocation as a kind of privilege, a kind of freedom:

> Everything feels like tourism in a room filled with strangers where you too are a stranger. I am drifting, *derive*, nearly anonymous, and in a state of mind conducive to corners and walls, where I stand alone, almost happy for the way I can spy and won't be seen, will remain unrecognized. Undisturbed I luxuriate, watching others dancing. Maybe noticed as the young American woman against the wall, if anything at all. (159)

Still later, she muses, "I'm not an exception in the world. Not an exception to its laws. Though hotel rooms provide the luxury, the privilege of seeming outside of them. In a detached home away from home, I can be captive to my thoughts. Easily captivated" (175). Such notions, that disorientation, dislocation, and detachment are conditions to revel in, are in and of themselves clear markers of the first-world tourist's privileged status—as a comparison with *Jasmine* immediately makes clear. For the undocumented immigrants that populate Mukherjee's novel, one must desperately try to fit and blend in to survive. For these immigrants, the dislocation and disorientation produced by the entry into the foreign is anything but exhilarating.

TRAVELER OR TOURIST?

As was the case for *The Sheltering Sky*, one of this novel's touchstones, the attempt to distinguish between traveler and tourist is a recurring concern. It may be helpful to recall that Bowles suggested that the difference between the two has to do with both movement abroad and the attitude one adopts while engaged in that movement. A true traveler, he claimed, is a nomad, moving from place to place over a period of years, never settling down. But, a traveler, unlike a tourist, is also willing to remain open intellectually to other cultures—other ways of acting, thinking, being. Echoes of these ideas appear in the following passage from *Motion Sickness* but in a way that conflates traveler and tourist, challenging Bowles's distinguishing criteria between the two:

> I don't really feel like staying or leaving. I'm not attached to this place, but place is unimportant to a traveler, if that's what I can be called. If it were important, people couldn't bear to move on. Jessica's favorite Gurdjieff quote, "If you love books don't become a bookseller," could be converted to "If you love places, don't leave home." When I get home I'll be made to reckon for my lack of curiosity, or I'll regret the untold opportunities I might have had. It's oppressive to be under inchoate directives to capture experience, to remember and save what's necessarily fleeting, even if it's been there for hundreds of years. Capture it quickly, imprint it in memory, try never to forget, have something to carry home. My souvenirs are mental and sentimental. That goes on cards to Steve and Charlotte. (93–94)

Although Tillman's narrator does appear to lead a nomadic life along the lines that Bowles suggests would mark her as a traveler, here she seems caught between staying and leaving, and her professed lack of curiosity would categorize her as a tourist in Bowles's view. Nevertheless, she resents the oppressive, inchoate directives, as she astutely describes them, that Dean MacCannell argues govern the activities of the modern tourist: itineraries must be planned and followed, and they must be built around particular tourist sites. When these sites are visited, the tourist must record his or her travel experience as evidence that they were genuine and authentic—thus the emphasis on photography and the purchase of souvenirs. Tillman's narrator seems torn between the dictates of travel and tourism. She clearly resents the oppressive mandates associated with tourism, yet she does a fair bit of sightseeing and is constantly buying and collecting postcards, which she describes as "playing cards that mark presence and absence" (88), suggesting the power of such mementos and souvenirs to authenticate travel experiences. Significantly, she rarely sends any of the postcards she writes and often destroys them after writing on them, which would seem to support her claim that their purpose is more sentimental than documentary. The destruction of the cards is perhaps an act of rebellion against those inchoate tourist directives—she prefers the "mental souvenirs" of her memories of places visited, even if she lacks physical proof that the visit took place. In fact, the final chapter of *Motion Sickness* is a compendium of the postcards she's written to various friends and fellow travelers, and the final lines of the novel are "I tear up the [postcard] that was meant for him. I may mail the others. I'm not sure" (204). Indecisiveness, indeterminacy, and impulsiveness govern her actions, and she herself questions whether she is a traveler, whether "that's what I can be called." For MacCannell, her ambivalence and indecisiveness are very much a product of postindustrial society, symptoms of a postmodern anxiety, which is the propulsive force of modern tourism. As I discussed in my chapter on Bowles, MacCannell theorized that tourism is a response to a pervasive fear that modern life is increasingly inauthentic or profane and is, above all, a quest for the authentic or sacred, driven by nostalgia. I suspect it is this anxiety with the way that modernization is changing places that prompted Bowles to lament to Tillman that place seems unimportant now.

MacCannell's diagnosis of postmodern Western society seems particularly well-suited for the world Tillman's narrator encounters in her travels. In Budapest, for example, while she argues politics with her lover and watches Hungarians enact "to perfection my expectation of this reality, this place, where newspapers hang from wooden sticks and the aroma of pastry is palpable," she, nevertheless, feels let down: "I can feel entirely indifferent to the content of what I say. A great postindustrial capitalist ennui engulfs me and sweeps away vestiges of involvement. Leaves me passionless and dissatisfied and incapable of movement. I'm threatened by this constantly" (102).

As her travels continue, this angst resurfaces in connection with her American identity, which she fears she will never entirely be able to shed, even as an expatriate:

> Maybe I'm as trapped as Pete. Although I may be in a melodrama, not a tragedy, white middle-class woman, from yet another dying empire, who dreams that change is possible but will remain unaware always of her fate, because she believes there is no fate, no history she can be tied to, so that her destiny will hinge on the shifts around her that she will call lucky or unlucky, personal triumph or defeat. National identity is like armor. On permanent loan from a museum. It's dull armor that I clink around in. Could I get an operation that would make me oblivious to symbols? (127)

This anxiety over being trapped within a cultural narrative that she cannot escape reappears in a later scene, as she gazes at herself in a mirror: "I may emit naiveté and hope in a limited way, the grand narrative I'm thrust out of overwhelming my individual predilections and deviations. The mirror over the hotel dresser offers no relief, no clue to my role in the larger story, one that is not easily framed and can't be studied objectively, can't take the country out of the girl when the girl is out of the country" (188–89). Interestingly, another American the narrator encounters accuses her of a kind of national betrayal, alleging that to become an expatriate is to betray one's country (174), yet the passages above suggest that one cannot discard national identity quite so easily.[3]

In her interview with Nicholls, Tillman acknowledged that the constraints of identity writ large, and national identity in particular, were very much on her mind as she wrote *Motion Sickness*: "All my books are in a way about limits, and about fighting those limits. . . . with *Motion Sickness* it was national identity and nationalism" (276). She elaborated on these points in her essay "Critical Fiction/Critical Self": "It's certainly on my agenda—to challenge the complacent, to question national, familial, racial, and sexual arrangements, to resist structures and institutions that serve the powerful and perpetuate powerlessness" (20), adding "So it's probably not surprising, given my concern with essentialisms of all kinds (and the construction of identity generally), that questioning national identity—how the nation state is inscribed in our different psyches, how each of us may be a repository of a national history and culture, how identity is inflected by nationality—that this was the impetus, even the mission (maybe impossible) for *Motion Sickness*, a novel whose narrator is a young, disquieted American woman traveling in Europe" (24). In her excellent analysis of Tillman's novel, Sue-Im Lee concurs, arguing that the narrator displays a deep suspicion of social norms and established cultural roles because on some level she senses that they mask oppressive (or, to use Tillman's term, limiting) ideas and institutions and are inescapable: "Throughout her travels, she watches her individual identity performing various roles in numerous ready-made plots of race, gender, and national identity" and has "a myriad of epistemological frameworks trained upon her. Recognition, in these scenes is nothing but a repetition of preexisting characterizations, plot typologies, and tensions" (148). "Tillman's focus on scenes of foreign encounter," Lee concludes, "highlights intersubjective recognition as being fundamentally oppressive, as being exchanges within a petrified sign system. . . . She is a subject who knows that she's consigned to being a reader, never a knower, and that as she's reading through the discourse of others, she's being read just as reductively, partially, and tautologically" (148–49).

If Tillman's narrator is unable to shed her national identity while traveling, so, too, is she unable to move outside of the touristic world she inhabits. Despite her nomadic, peripatetic wanderings, it's worth noting that she is dependent upon the infrastructure we associate with tourists and well-to-do travelers. This is no twenty-something, adventuresome college student staying at hostels and hitchhiking across the continent, nor is she Kit Moresby, running off recklessly into the Sahara in an attempt to escape her horror of the modern world. Rather, she always stays in good hotels, eats at nice restaurants, and visits the expected tourist sites. Moreover, the actual act of travel from place to place is, with a few exceptions, conspicuously absent in her account, an indication of how those of economic means take their mobility for granted; the fact that our narrator thinks this movement from point A to point B unremarkable is itself a reminder of her privilege.

These differences come into stark relief in one of the novel's most interesting chapters, describing the narrator's stay in Tangier, Morocco. Tangier, of course, was also Bowles's adopted home, and the chapter is filled with allusions to *The Sheltering Sky*.[4] It opens with a remarkable homage to famous women travelers (both real and fictional) who came to Africa in the nineteenth and twentieth centuries in search of freedom and adventure, among them Isak Dinesen, Freya Stark, Isabelle Eberhardt, and Kit Moresby. Tillman's narrator describes these women as "remarkably resilient" and dedicated, above all, to the lure of adventure and the unknown—sacrificing all that they have, and enduring scorn, poverty, madness, and even loss of self, to pursue these goals. As she contemplates the enormity of those sacrifices, the narrator muses:

> She is eager to begin her trip. The future was the desert, as the future now is outer space, a future that science-fiction writers long ago colonized and have been waiting impatiently for us to enter. Perhaps going into the desert would be going backward since time might be curved, and anything can be the present, past or future. If this is true then my not going into the desert—to lose myself, my mind—is not just a matter of cowardice or lack of adventurousness. I may have already been there or will be going there. Still, I think about it often. Losing myself in the desert. Abandoning everything, being abandoned. Taking up residence in a small town in Iowa where no one would know me. I'd cut off all ties to my past, I'd use the frontier, I'd become a frontier. I'd be known as that woman at the edge of the real last frontier. I'd have a front porch in my frontier version of reality. I'd sit on that front porch and gaze out at limitless skies and pastures. Relativity would be as close as I came to relatives. Not that I have many, just my mother, really. I owe her a letter. (117)

Several things in this remarkable passage are worth noting. As Heidi Wallace has observed, in suggesting that science fiction has already predetermined our notions of outer space (and even the future), Tillman is also calling our attention to the ways in which travel literature over-determines our notions of such famous destinations as Tangier and the Sahara Desert, imbuing them with exotic and mythic qualities.[5] The intertextuality of the passage is also striking. Not only does Tillman allude to Kit Moresby, but also the narrator's musings about losing herself in a small Iowa town and gazing out at the frontier seem to me a tantalizing reference to Jyoti of *Jasmine*, who fled to Iowa and tried to cut off all ties to her past to start a new life. Not coincidentally, Mukherjee's novel ends with Jyoti leaving Iowa for California and invoking the frontier as Tillman's narrator does, as she eagerly looks forward to the possibilities of a new life: "Adventure, risk, transformation: the frontier is pushing indoors through uncaulked windows. . . . I am out the door and in the potholed and rutted driveway . . . greedy with wants and reckless from hope" (240–41). And, although Tillman's narrator seems to harbor a similar longing to venture into the unknown, she is hemmed in by fear and caution, and the abrupt dissolution of her fantasy into thoughts of her mother and daughterly duty suggest that she is unlikely to ever venture off the beaten path. While she admires her pantheon of solitary women travelers and adventurers, and clings to a faint hope that someday she might follow in their footsteps, this possibility seems unlikely and again underscores her status as a tourist, not a traveler. Moreover, the series of literary allusions contained in this chapter's opening pages highlights the intertextual nature of travel literature, and Tillman's awareness of writing within, and in response to, that tradition.[6] She makes us recognize the influence of travel fictions upon travelers and how they shape what we as travelers expect to experience: our measure of authenticity. Tillman's narrator longs for a kind of "true tourism," which she describes as being able to see oneself as others do (120), and this desire is, I think, further evidence of her awareness of her economic privilege and the discomfort it produces. For although she wishes she could learn what the natives of the host countries she visits think of her, she desperately does not want to be seen as an imperialist nor as an American—she does not want to be tied to her national identity (124–27). And yet the fact that she is inescapably a product of, and implicated in, an exploitative, capitalist system is underscored just a few moments later, as she describes the Moroccan women who work as maids in her hotel as "opaque," a description that underscores the lack of visibility and taken-for-granted status of the underpaid workers who support the privileged travelers on their grand tours. Notably, this is one of the few times that she mentions the workers who make her travels possible, the "women who cleaned my room and whom I never really saw" (130).

The narrator's longing for a true tourism, an opportunity to get outside one's body and see oneself as others do, is akin to what Sue-Im Lee calls "recognition as knowing afresh" and what Paul Fussell would describe as one of the things that distinguish the traveler from the tourist. In other words, the traveler sees afresh, sees anew, because she is open to new possibilities, whereas the tourist sees only what has been prepared for her by the arts of mass publicity (as Fussell puts it). Lee might respond, I think, that travel as Fussell conceives of it is not possible, only tourism; because we are trapped in the grand narratives and ideologies we are interpellated by, travel takes place in an epistemological dead-end. In her interview with Nicholls, Tillman observes, "Obviously I'm being ironic about calling it a true tourism. It's not that you find yourself, but that you can look at yourself. When you go somewhere else and see new things you're also trying to be a different person in a new place. *Of course you can't ever do that really*. But what you keep hoping for is getting outside yourself, your country . . . so that in a way you wouldn't exist. That may be a positive form of tourism" (272, emphasis added).

A LAND MADE OF ITS REPRESENTATIONS

While viewing fantastical paintings by Uccello at the National Gallery in London, the narrator imagines "a land made of its representations" that can be approached but never entered (199). The phrase is suggestive of the vexed relationship between the traveler and the places he or she visits. Like Bowles, MacCannell, and Jonathan Culler, Tillman suggests that famous tourist destinations like New York, Paris, and Tangier have become so over-determined by their endless representation in various forms that they have become unknowable and inaccessible in the way that MacCannell's archetypal tourist desires to know and access them:

> I tell Arlette New York is like Paris. That's probably not so. Days turn into nights here the way they do in New York. The cities never sleep. Cities don't sleep. Apart from friends and my block, or the skyline, as familiar and remote as Marilyn Monroe's figure, different neighborhoods are convincing real-life locations for movies I've seen. Or the city is posters, a backdrop of images and words. . . . I don't really see New York. Or when I see it, it's always the same. I can't see it. Or for that matter America. Just the way I don't see Paris. Or Europe. It's inaccessible to tourism, or it's all tourism. (150)

That the narrator is a New Yorker suggests that the influence of representations and simulacra of famous places is so powerful it even affects the perception of someone who has lived there all her life. It's not a coincidence, then, that as she travels, the narrator is continually relating to the people, cultures, and places she visits through travel fantasies that are shot through with references from books, movies, and other forms of popular culture:

> I'd be sitting at a café, a famous one, because I wouldn't be easily discovered in an out of the way place, though Sylvie found me at one. I'd be sitting at a café at which I was already not a newcomer, and there would be a glass of Pernod on the small round table in front of me. I'd have a book in my hand. But I'd be observing the couple, the lovers, near me, arguing in an excited way, about something I wouldn't comprehend. They would be characters from an early Godard film or a more recent Bresson. If they were from *Breathless*, I might understand them, with Jean Seberg speaking an American French slow enough even for me. (157)

As Lee observes, "As [the narrator] comprehends the foreign setting and people by anchoring them to the plots, scenarios, settings, and characters from her archive of movies, musicals and books, her mind's wanderings represent a subjectivity dominated by ready-made signs of a postmodern image culture" (148).

Of course, it is not only the influence of earlier narratives and representations which affect a travel writer's perception of people, places, and cultures and his or her subsequent attempt to represent it in a narrative of their own but also the problem of memory and selectivity. As she tries to recount her recent travels to a fellow American expatriate, the narrator reflects upon precisely this issue: "There's so much forgotten or not chosen. Some memories are always the bridesmaid and never the bride" (173). The narrator's obsession with assigning the right postcard to the right person is akin to how travelers carefully construct narratives around their experiences—making careful choices about what to include or exclude. The postcards she writes are full of indecision (I may do this, I may do that), and she doesn't send them because to do so would be to commit, to decide, and she prefers to live in a "state of flux," as one of the chapter headings calls it, shuttling back and forth between European cities. As Tillman herself puts it, "The question of agency haunts the novel" (*The Broad Picture*, 20).

Tillman also provides a perceptive critique of travel guidebooks, a genre that is a key component of the tourism industry. As Tillman's narrator astutely observes, these guidebooks are carefully constructed narratives that often bear little resemblance or relationship to the people, places, and cultures they purport to describe:

> I switch from Italian grammar to a guidebook describing the streets I just walked. Alfred's travel books remind me of cookbooks. They must be written by people with terrific imaginations who can sit down at a desk, relive sensuous experiences, or if not relive them, fantasize wildly, then pluck words that satisfy from a storehouse of memories, a warehouse of words that match sensations. Such flights of fancy attempt to occupy the empty space of someone else's imagination, where they're meant to dress up an otherwise colorless existence

> or inflame the senses like pornography. The descriptions don't make me see better; they're more like putting on 3-D glasses which perch shakily on the nose, always in danger of falling off, or are something entirely different from the object described, to be appreciated for themselves only. Which I suppose is OK and makes cookbooks and travel guides akin to science fiction. But everything sparkles too much, is too luscious or exquisite. Leaves a nasty aftertaste, like depression. (59)

The narrator's observations about tourist guidebooks could easily be applied to many travel narratives and travel fictions, as well. As I have discussed throughout this book, nonfiction travel narratives often distort or exaggerate—for a whole host of reasons and motives—in ways that do not allow the reader "see better." Rather, as Tillman's narrator suggests in the previous passage, one of the ways travel literature distorts, as David Spurr has persuasively argued, is by aestheticizing other places and cultures, making them more beautiful—"sparkling, luscious, and exquisite."[7] It is this slippage between the descriptions of people and places in travel literature and the places and cultures themselves which leads to the common tourist angst discussed at the outset of this chapter—the tourist's sense that places and cultures do not correspond with their representations and that reality, itself, is somehow inauthentic.

An interesting opposing view to consider is put forward by Paul Fussell in *Abroad*, in which he contends that not all travel books are created equal. Fussell is careful to distinguish between a travel book, which he argues is really a collection of learned essays couched in a travel memoir, and travel guidebooks, which offer advice and tips, and thus serve a more utilitarian purpose. In Fussell's view, true travel books are at their core displaced quest and pastoral romances: "As in a [quest] romance, the modern traveler leaves the familiar and predictable to wander, episodically, into the unfamiliar or unknown, encountering strange adventures, and finally, after travail and ordeals, returns safely. Somehow, we feel a travel book isn't wholly satisfying unless the traveler returns to his starting point: the action, as in a quest romance, must be completed" (208). In this regard, Tillman's novel is not a typical travel book.

Moreover, Fussell continues, travel books are often infused with a nostalgia for a mythic past (the preindustrial past, MacCannell would say), which causes them to function as displaced pastoral romances, frequently depicting "a beautiful relationship between rich and poor," a phrase Fussell borrows from William Empson. It is precisely this pastoral quality, I hasten to add, which is one of travel literature's most problematic tendencies, whether fiction or nonfiction. As Kincaid, Said, Pratt, and a plethora of other scholars have noted, the frequent depiction of poverty as picturesque and local peoples as the happy servants of privileged first-world travelers is one of travel writing's most damaging tendencies. Thus, while both Fussell and Tillman recognize the aestheticizing power of much travel literature, for Fussell, this tendency is a great virtue, whereas Tillman views it (correctly, I think) as more problematic.

One of the chapter epigraphs in *Motion Sickness* is taken from Julia Kristeva, who wrote: "The expatriate represents, in fact, the normal state of an average citizen in this last part of the 20th century," reminding us of the frequent, forced displacement which seems so characteristic of life since World War II. As I have tried to demonstrate here, *Motion Sickness* is a novel that examines the privileges of the first-world traveler in order to highlight the enormous differences between tourist/expatriate and refugee/exile. Tillman remarked to Nichols that, as she wrote the novel, she was not only thinking about those exiled *from* their countries but also those exiled *within* their countries as marginalized populations, through various forms of disenfranchisement: "I wanted to turn it all around and say, OK, here's this travel business, but you can think about this differently. You can think that where you are is also not a secure place to be, and that you're maybe feeling as uprooted as somebody who's not in their own country, I mean, think about all the different populations in America who aren't exactly served by that system" (271).

CONCLUDING REMARKS

Like the other works of fiction discussed in this study, Tillman's *Motion Sickness* displays a remarkable awareness of operating within a specific literary tradition and it engages in a productive dialogue with other works in that tradition, often calling attention to problematic aspects of travel fictions. As discussed in this chapter, for example, Tillman calls the reader's attention to the privileges of the white, first-world traveler and how, for many other people in the world, travel is not a choice but instead a dangerous situation into which they are thrust and to which they must find a way to survive. And Tillman critiques the tendency of travel fictions to predetermine and overdetermine our notions of people and places to such a degree that contradictory first-hand experiences are often shrugged off by the traveler as wrong or, at the very least, disappointing. As she puts it so well, travel literature (and other media) has transformed the world into a land made of its representations, a land that we can approach but never enter, no matter how much we might desire to do so.

NOTES

1. Tillman has said that she intended this first chapter to function as a kind of movie trailer for the upcoming feature, reinforcing the notion that the opening chapter functions metonymically.

2. See Patrick McGrath, "Lynne Tillman," *BOMB* 35 (Spring 1991): n. pag.

3. Tillman echoed this sentiment regarding what she perceives to be a distrust of expats by Americans in her interview with Nicholls. See page 272.

4. The owner of her hotel in Tangier, for example, is Mr. Mrabet, perhaps an allusion to Mohammed Mrabet, one of Morocco's most famous authors and a friend of Bowles, who translated his work into English.

5. Heidi Wallace, unpublished manuscript, 2014.

6. Lee also notes the intertextuality of this chapter but interprets it as a self-conscious parody of the iconic "wandering woman," which impresses upon the reader how "congealed" and cliché the archetype has become. See Lee, 148.

7. See David Spurr, *The Rhetoric of Empire: Colonial Discourse in Journalism, Travel Writing, and Imperial Administration*, Durham, NC: Duke University Press, 1993.

Conclusion

As I worked on this book over the years, I often thought of the memorable opening line to L. P. Hartley's novel *The Go-Between* (1953): "The past is a foreign country: they do things differently there"—a reminder that our relationship to the past is always mediated by narrative and, therefore, our recounting of past events, including journeys to other places and our interactions with other people, is always distorted to some degree (17). That Hartley chose a geographical metaphor to express this idea only underscores how closely associated travel writing is with otherness. Like travelers, the readers of travel literature expect to encounter the Other. We expect other places to be different than our own, and we expect the people who live there to behave differently than those of our own society, our own culture. Unfortunately, this expectation of otherness often goes hand in hand with a dehumanization, with an inability to see one's shared humanity beyond cultural differences, as, for example, in the travel writing of John Stephens or in the parody of such narratives of exploration by Clarice Lispector.

The power of travel literature to shape our expectations and perception of other cultures is not insignificant. In Don DeLillo's travel fiction *The Names*, his globe-trotting protagonist claims that he "goes everywhere twice, once to get the wrong impression and once to strengthen it" (255). The tongue-in-cheek admission underscores how powerfully travel writing and travel fictions shape our views of other people and places. As Fussell observes in *Abroad*, travel and literature are inextricably linked and travel literature has such a hold over our imagination that we are drawn to its representations and stereotypes, even when they are challenged by our own lived experience. When we travel, we don't want to find reality; we want to find the iconic and the exotic. Like Tillman's narrator in *Motion Sickness*, we desire to live in a world of representations, rather than the actual one we encounter on our journeys. That my examples of these insights into travel writing are taken from three novels is yet one more vivid reminder of the vitally important critical work performed by travel fictions, an aspect of travel literature that is so often neglected and overlooked, and which I have tried to remedy here. They provide thoughtful, theoretical critiques of different kinds of travel, as we saw in *The Sheltering Sky*, *Volkswagen Blues*, *Jasmine*, and *Motion Sickness*. And while some travel narratives, like that of Stephens, present a case study in how travel literature has often served colonial and capitalist aggression and acquisition, the genre has also been used to provide important socioeconomic and cultural critiques, as in Jack London's *The Road* and Jamaica Kincaid's *A Small Place*. In sum, if we are willing to revisit this fascinating genre with fresh eyes and with an awareness of how subversive and sophisticated both travel narratives and travel fictions can be, we will be rewarded intellectually many times over and avoid getting and strengthening the wrong impression about them, even though wrong impressions may be what we both desire and expect.

Bibliography

Adams, Rachel. *Continental Divides: Remapping the Cultures of North America.* Chicago: University of Chicago Press, 2009.

Adler, Judith. "Youth on the Road: Reflections on the History of Tramping." *Annals of Tourism Research* 12 (1985): 335–54.

Bahri, Deepika. "Always Becoming: Narratives of Nation and Self in Bharati Mukherjee's *Jasmine*." *Women, America, and Movement: Narratives of Relocation.* Ed. Susan L. Roberson. Columbia, MO: University of Missouri Press, 1998.

Baudrillard, Jean. *Selected Writings.* Ed. Mark Poster. Stanford, CT: Stanford University Press, 1988.

Benjamin, Walter. "The Work of Art in the Age of Mechanical Reproduction." *Illuminations.* Ed. Hannah Arendt. 1968. New York: Schocken, 1985. 217–51.

Bowen, Roger. "Journey's End: Conrad as Revenant in Alex Garland's *The Beach*." *Conradiana* 39.1 (2007): 39–57.

Bowles, Paul. *The Sheltering Sky.* New York: New Directions, 1949.

———. *Without Stopping.* New York: G. P. Putnam's Sons, 1972.

———. *Their Heads Are Green and Their Hands Are Blue: Scenes from the Non-Christian World.* 1963. Hopewell, NJ: Ecco, 1984.

———. "A Distant Episode." *The Delicate Prey and Other Stories.* 1950. Hopewell, NJ: Ecco, 1995. 290–307.

Bowles, Jane. *Two Serious Ladies.* 1943. New York: Ecco, 2014.

Brickhouse, Anna. *Transamerican Literary Relations and the Nineteenth-Century Public Sphere.* Cambridge: Cambridge University Press, 2004.

Burkhart, Matt. "Rewriting the West(ern): *Shane*, Jane, and Agricultural Change in Bharati Mukherjee's *Jasmine*." *Western American Literature* 43.1 (Spring 2008): 5–22.

Cabañas, Miguel A. *The Cultural "Other" in Nineteenth-Century Travel Narratives: How the United States and Latin America Described Each Other.* Lewiston, NY: Edwin Mellen Press, 2008.

Calderón de la Barca, Frances. *Life in Mexico.* 1842. Berkeley, CA: University of California Press, 1982.

Carter-Sanborn, Kristin. "'We Murder Who We Were': *Jasmine* and the Violence of Identity." *American Literature* 66.3 (1994): 573–93.

Clendinnen, Inga. "'Fierce and Unnatural Cruelty': Cortes and the Conquest of Mexico." *Representations* 33 (1991): 65–100.

Clifford, James. *Routes: Travel and Translation in the Late Twentieth Century*. Cambridge, MA: Harvard UP, 1997.

Coe, Andrew. *Chop Suey: A Cultural History of Chinese Food in the United States*. Oxford: Oxford University Press, 2009.

Cooper Alarcón, Daniel. *The Aztec Palimpsest: Mexico in the Modern Imagination*. Tucson, AZ: University of Arizona Press, 1997.

Cresswell, Tim. *The Tramp in America*. London: Reaktion, 2001.

Culler, Jonathan. "Semiotics of Tourism." *American Journal of Semiotics* 1.1–2 (1981): 127–40.

DeLillo, Don. *The Names*. New York: Random House, 1989.

DePastino, Todd. *Citizen Hobo: How a Century of Homelessness Shaped America*. Chicago: University of Chicago Press, 2003.

———. "Introduction." *The Road*. New Brunswick, NJ: Rutgers University Press, 2006.

Dlaska, Andrea. *Ways of Belonging: The Making of New Americans in the Fiction of Bharati Mukherjee*. Austria: Braumuller, 1999.

Donnell, Alison. "She Ties Her Tongue: The Problems of Cultural Paralysis in Postcolonial Criticism." *ARIEL* 26.1 (1995): 101–16.

Durica, Paul. "Flux Americana: The Tramp and the Problem of Mobility in American Literature, 1870–1960." *Dissertation Abstracts International*. May 2014.

Edlmair, Barbara. *Rewriting History: Alternative Versions of the Caribbean Past in Michele Cliff, Rosario Ferre, Jamaica Kincaid, and Daniel Maximin*. Vienna: Braumuller, 1999.

Edwards, Brian T. "Sheltering Screens: Paul Bowles and Foreign Relations." *American Literary History* 17.2 (2005): 307–34.

Edwards, Justin. *Understanding Jamaica Kincaid*. Columbia, SC: University of South Carolina Press, 2007.

Etulain, Richard, ed. *Jack London on the Road: The Tramp Diary and Other Hobo Writings*. Logan, UT: Utah State University Press, 1978.

Evans, R. Tripp. *Romancing the Maya: Mexican Antiquity in the American Imagination, 1820–1915*. Austin, TX: University of Texas Press, 2004.

Ewert, Jeanne. "'Great Plant Appropriators' and Acquisitive Gardeners: Jamaica Kincaid's Ambivalent Garden (Book)." *Jamaica Kincaid and Caribbean Double Crossings*. Ed. Linda Lang-Peralta. Newark, NJ: University of Delaware Press, 2006.

Feied, Frederick. *No Pie in the Sky: The Hobo as American Cultural Hero in the Works of Jack London, John Dos Passos, and Jack Kerouac*. New York: Citadel, 1964.

Fussell, Paul. *Abroad: British Literary Traveling Between the Wars*. Oxford: Oxford University Press, 1980.

Graham-Bartolini, Alison. "The Advantage of Estrangement in Mukherjee's *Jasmine*." *Indian Writers: Transnationalisms and Diasporas*. Eds. Jaspal K. Singh and Rajendra Chetty. New York: Peter Lang, 2010.

Grewal, Inderpal. *Transnational America*. Durham, NC: Duke University Pres, 2005.

Guenther, Beatrice. "Land and Cityscape in Lise Gauvin's and Jacques Poulin's Narratives: between Cultural Memory and *L'Invention du Quotidien*." *Land and Landscape in Francographic Literature: Remapping Uncertain Territories*. Eds. Magali Compan and Katarzyna Pieprzak. Newcastle: Cambridge Scholars Publishing, 2007. 60–87.

Haggard, H. Rider. *Montezuma's Daughter*. London: Longmans, 1893.

Harte, Bret. "My Friend, the Tramp." 1877. *Drift from Two Shores*. Cambridge, MA: Houghton, 1878.

Hartley, L. P. *The Go-Between*. 1953. New York: New York Review Books, 2002.

Harvey, Bruce A. *American Geographics: U.S. National Narratives and the Representation of the Non-European World, 1830–1865*. Stanford, CT: Stanford University Press, 2001.

Hibbard, Allen. *Paul Bowles: A Study of the Short Fiction*. New York: Twayne, 1993.

———. *Paul Bowles: Magic and Morocco*. San Francisco: Cadmus, 2004.

Hyman, Roger. "Writing against knowing, writing against certainty, or What's really under the veranda in Jacques Poulin's *Volkswagen Blues*." *Journal of Canadian Studies* 34.3 (1999): 106–33.

Ippolito, Emilia. *Caribbean Women Writers: Identity and Gender*. Rochester, NY: Camden House, 2000.

Johnson, David E. "'Writing in the Dark': The Political Fictions of American Travel Writing." *American Literary History* 7.1 (1995): 1–27.

Kaplan, Caren. *Questions of Travel: Postmodern Discourses of Displacement*. Durham, NC: Duke University Press, 1996.

Kincaid, Jamaica. *A Small Place*. New York: Plume, 1989.

———. *My Garden (Book)*. New York: Farrar, Straus, and Giroux, 1999.

Leask, Nigel. "A Yankee in Yucatán: John Lloyd Stephens and the Lost Cities of America." *Travel Writing in the Nineteenth Century: Filling the Blank Spaces*. Ed. Tim Youngs. London: Anthem Press, 2006.

Lee, Sue-Im. "Recognition as a Depleted Source in Lynne Tillman's Motion Sickness." *Symploke* 12.1–2 (2004): 139–51.

———. *A Body of Individuals: The Paradox of Community in Contemporary Fiction*. Columbus, OH: Ohio State University Press, 2009.

Lennon, John. *Boxcar Politics: The Hobo in U.S. Culture and Literature, 1869–1956*. Boston: University of Massachusetts Press, 2014.

Lenz, William. "Identity in John Lloyd Stephens's *Incidents of Travel in Central America, Chiapas, and Yucatán*." In *Travel Culture: Essays on What Makes Us Go*. Ed. Carol Traynor Williams, 79–87. Westport, CT: Praeger, 1998.

———. *Ruins, Revolution, and Manifest Destiny: John Lloyd Stephens Creates the Maya*. New York: Peter Lang, 2013.

Levin, Stephen. *The Contemporary Anglophone Travel Novel: The Aesthetics of Self-Fashioning in the Era of Globalization*. New York: Routledge, 2008.

Lispector, Clarice. "The Smallest Woman in the World." *The Eye of the Heart*. Ed. Barbara Howes. New York: Avon/Bard, 1974. 447–54.

London, Jack. *The Road*. 1907. Ed. Todd DePastino. New Brunswick, NJ: Rutgers University Press, 2006.

MacCannell, Dean. *The Tourist: A New Theory of the Leisure Class*. 1976. New York: Random, 1989.

MacDonald-Smythe, Antonia. *Making Homes in the West/Indies: Constructions of Subjectivity in the Writings of Michelle Cliff and Jamaica Kincaid.* New York: Garland, 2001.

MacFarlane, Heather. "*Volkswagen Blues* Twenty-Five Years Later: Revisiting Poulin's Pitsémine." *Studies in Canadian Literature* (2009): 1–21.

Macpherson, Heidi Slettedahl. *Transatlantic Women's Literature*. Edinburgh: Edinburgh University Press, 2008.

Martino, Andrew. "The Vanishing Point: the Dis-Integration of Female Identity in Paul Bowles's *The Sheltering Sky*." *South Atlantic Review* 71.2 (2006): 87–114.

McGrath, Patrick. "Lynne Tillman." *BOMB* 35 (Spring 1991): n. pag.

Mihn-ha, Trinh T. "Not You/Like You: Post-Colonial Women and the Interlocking Questions of Identity and Difference." *Making Face, Making Soul: Haciendo Caras.* Ed. Gloria Anzaldua. San Francisco: Aunt Lute, 1990. 371–75.

Mikkonen, Kai. *Narrative Paths: African Travel in Modern Fiction and Nonfiction.* Columbus, OH: Ohio State University Press, 2015.

Miraglia, Anne Maria. "Texts Engendering Texts: A Québecois Rewriting of American Novels." *Intertextuality in Literature and Film*. Eds. Elaine Cancalon and Antoine Spacagna. Gainesville: University of Florida Press, 1988. 49–60.

Mukherjee, Bharati. *Jasmine*. New York: Grove Press, 1989.

Mullins, Greg. *Colonial Affairs: Bowles, Burroughs, and Chester Write Tangier.* Madison, WI: University of Wisconsin Press, 2002.

Nicholls, Peter. "A Conversation with Lynne Tillman." *Textual Practice* 9.2 (1995): 269–84.

Nissen, Axel. "A Tramp at Home: 'Huckleberry Finn,' Romantic Friendship, and the Homeless Man." *Nineteenth-Century Literature* 60.1 (2005): 57–86.

Papayanis, Marilyn Adler. *Writing in the Margins: The Ethics of Expatriation from Lawrence to Ondaatje*. Nashville, TN: Vanderbilt UP, 2005.

Patteson, Richard F. *A World Outside: The Fiction of Paul Bowles*. Austin, TX: University of Texas Press, 1987.

Paul Singh, Ryan. "'I want to be surprised when I hear your voice': Who Speaks for *Jasmine*?" *Indian Writers: Transnationalisms and Diasporas*. Eds. Jaspal K. Singh and Rajendra Chetty. New York: Peter Lang, 2010.

Peixoto, Marta. *Passionate Fictions: Gender, Narrative, and Violence in Clarice Lispector.* Minneapolis, MN: University of Minnesota Press, 1994.

Phillips, Caryl. *A New World Order*. New York: Random House, 2002.

Photinos, Christine. "The Figure of the Tramp in Gilded Age Success Narratives." *The Journal of Popular Culture* 40.6 (2007a): 994–1018.

———. "Tracking Changes in Jack London's Representation of the Railroad Tramp." *The Journal of American Culture* 30.2 (2007b): 175–86.

Poulin, Jacques. *Volkswagen Blues*. Trans. Sheila Fischman. 1984. Toronto: Cormorant, 2004.

Pratt, Mary Louise. *Imperial Eyes: Travel Writing and Transculturation.* London: Routledge, 1992.

Prescott, W. H. *History of the Conquest of Mexico*. 1843. New York: Random, n.d.

Rosenberg, Judith. "Taking Her Measurements: Clarice Lispector and 'The Smallest Woman in the World.'" *Critique: Studies in Contemporary Fiction* 30.2 (Winter 1989): 71–76.

Rosenstreich, Susan. "God the Father or Mother Earth?: *Nouvelle France* in Two Québec Novels of the 1980s." *L'Esprit Createur* 48.1 (2008): 120–30.

Said, Edward. *Orientalism*. 1978. New York: Random, 1979.

Silko, Leslie Marmon. "Fences Against Freedom." *Yellow Woman and a Beauty of the Spirit*. New York: Simon and Schuster, 1996a. 100–14.

———. "The Border Patrol State." *Yellow Woman and a Beauty of the Spirit*. New York: Simon and Schuster, 1996b. 115–23.

Simpson, Mark. *Trafficking Subjects: The Politics of Mobility in Nineteenth-Century America*. Minneapolis, MN: University of Minnesota Press, 2005.

Sontag, Susan. *On Photography*. 1977. New York: Doubleday, 1989.

Spurr, David. *The Rhetoric of Empire: Colonial Discourse in Journalism, Travel Writing, and Imperial Administration*. Durham, NC: Duke University Press, 1993.

Stephens, John L. *Incidents of Travel in Central America, Chiapas, and Yucatán*. Vol. I and II. 1841. New York: Dover, 1969.

Stephens, John L. *Incidents of Travel in Yucatán*. Vols. I and II. 1843. New York: Dover, 1963.

Stewart, Lawrence. *Paul Bowles: The Illumination of North Africa*. Carbondale, IL: Southern Illinois University Press, 1974.

Swafford, Kevin. "Tramping with Jack London: Poverty, Performance, and Social Critique in *The Road*." *Journal of the Midwest Modern Language Association* 51.1 (2018): 173–95.

Tichi, Cecelia. "Canonizing Economic Crisis: Jack London's *The Road*." *American Literary History* 23.1 (2011): 19–31.

Tillman, Lynne. *Motion Sickness*. New York: Poseidon Press, 1991.

———. "Critical Fiction/Critical Self." *The Broad Picture*. London: Serpent's Tail, 1997: 18–25.

———. *What Would Lynne Tillman Do?* Brooklyn: Red Lemonade, 2014.

Urry, John. *The Tourist Gaze: Leisure and Travel in Contemporary Societies*. London: Sage, 1990.

Warhol, Robyn. "*Jasmine* Reconsidered: Narrative Structure and Multicultural Subjectivity." *Analyzing World Fiction: New Horizons in Narrative Theory*. Ed. Frederick Luis Aldama. Austin, TX: University of Texas Press, 2011.

Weisman, Adam. "Postcolonialism in North America: Imaginative Colonization in Henry David Thoreau's *A Yankee in Canada* and Jacques Poulin's *Volkswagen Blues*." *Massachusetts Review* 36.3 (1995): 477–500.

Weiss, Timothy. *Translating Orients: Between Ideology and Utopia*. Toronto: University of Toronto Press, 2004.

Williams, Claire. "More than Meets the Eye, or a Tree House of Her Own: A New Look at a Short Story by Clarice Lispector." *Portuguese Studies* 14 (1998): 170–80.

Index

About the Author

Daniel Cooper Alarcón is an associate professor of English at the University of Arizona in Tucson, where he has lived for more than thirty years. He is the author of *The Aztec Palimpsest: Mexico in the Modern Imagination* (1997) and numerous articles, including essays on Cormac McCarthy, Ana Castillo, José Antonio Villarreal, John Lloyd Stephens, and Paul Bowles.